Perchlorate Ecotoxicology

Other titles from the Society of Environmental Toxicology and Chemistry (SETAC)

Mercury Cycling in a Wetland-Dominated Ecosystem: A Multidisciplinary Study
O'Driscoll, Rencz, Lean
2005

Atrazine in North American Surface Waters: A Probabilistic Aquatic Ecological Risk Assessment
Giddings, editor
2005

Effects of Pesticides in the Field
Liess, Brown, Dohmen, Duquesne, Hart, Heimbach, Kreuger, Lagadic, Maund, Reinert, Streloke, Tarazona
2005

Human Pharmaceuticals: Assessing the Impacts on Aquatic Ecosystems
Williams, editor
2005

Toxicity of Dietborne Metals to Aquatic Organisms
Meyer, Adams, Brix, Luoma, Stubblefield, Wood, editors
2005

Toxicity Reduction and Toxicity Identification Evaluations for Effluents, Ambient Waters, and Other Aqueous Media
Norberg-King, Ausley, Burton, Goodfellow, Miller, Waller, editors
2005

Use of Sediment Quality Guidelines and Related Tools for the Assessment of Contaminated Sediments
Wenning, Batley, Ingersoll, Moore, editors
2005

Life-Cycle Assessment of Metals
Dubreuil, editor
2005

Working Environment in Life-Cycle Assessment
Poulsen and Jensen, editors
2005

For information about SETAC publications, including SETAC's international journals, *Environmental Toxicology and Chemistry* and *Integrated Environmental Assessment and Management*, contact the SETAC Administrative Office nearest you:

SETAC Office
1010 North 12th Avenue
Pensacola, FL 32501-3367 USA
T 850 469 1500 F 850 469 9778
E setac@setac.org

SETAC Office
Avenue de la Toison d'Or 67
B-1060 Brussels, Belgium
T 32 2 772 72 81 F 32 2 770 53 86
E setac@setaceu.org

www.setac.org

Environmental Quality Through Science®

Perchlorate Ecotoxicology

Edited by

Ronald J. Kendall
Philip N. Smith

Coordinating Editor of SETAC Books

Joseph W. Gorsuch
Gorsuch Environmental Management Services, Inc.

published by the society of environmental toxicology and chemistry (setac)

SETAC PRESS

Cover by Kerri Charlton

Library of Congress Cataloging-in-Publication Data

Perchlorate ecotoxicology / edited by Ronald J. Kendall and Philip N. Smith.
p. cm.
Includes bibliographical references and index.
ISBN 1-880611-90-2 (alk. paper)
1. Perchlorates--Environmental aspects. 2. Perchlorates--Toxicology. 3. Water--Purification--Perchlorate removal. I. Kendall, Ronald J. II. Smith, Philip N., 1965-

TD427.P33P456 2006
571.9'5--dc22 2006051194

ISBN-13: 978-1-880611-90-6

ISBN-10: 1-880611-90-2

Printed in the United States of America
10 09 08 07 06 10 9 8 7 6 5 4 3 2 1

♾ The paper used in this publication meets the minimum requirements of the American National Standard for Information Sciences—Permanence of Paper for Printed Library Materials, ANSI Z39.48-1984.

Reference Listing: Kendall RJ, Smith PN, editors. 2006. Perchlorate ecotoxicology. Pensacola (FL): Society of Environmental Toxicology and Chemistry (SETAC). 288 p.

SETAC Publications

Books published by the Society of Environmental Toxicology and Chemistry (SETAC) provide in-depth reviews and critical appraisals on scientific subjects relevant to understanding the impacts of chemicals and technology on the environment. The books explore topics reviewed and recommended by the Publications Advisory Council and approved by the SETAC North America, Latin America, or Asia/Pacific Board of Directors; the SETAC Europe Council; or the SETAC World Council for their importance, timeliness, and contribution to multidisciplinary approaches to solving environmental problems. The diversity and breadth of subjects covered in the series reflect the wide range of disciplines encompassed by environmental toxicology, environmental chemistry, and hazard and risk assessment, and life-cycle assessment. SETAC books attempt to present the reader with authoritative coverage of the literature, as well as paradigms, methodologies, and controversies; research needs; and new developments specific to the featured topics. The books are generally peer reviewed for SETAC by acknowledged experts.

SETAC publications, which include Technical Issue Papers (TIPs), workshop summaries, newsletter (*SETAC Globe*), and journals (*Environmental Toxicology and Chemistry* and *Integrated Environmental Assessment and Management*), are useful to environmental scientists in research, research management, chemical manufacturing and regulation, risk assessment, and education, as well as to students considering or preparing for careers in these areas. The publications provide information for keeping abreast of recent developments in familiar subject areas and for rapid introduction to principles and approaches in new subject areas.

SETAC recognizes and thanks the past coordinating editors of SETAC books:

Andrew Green, International Zinc Association
Durham, North Carolina, USA

C.G. Ingersoll, Columbia Environmental Research Center
US Geological Survey, Columbia, Missouri, USA

T.W. La Point, Institute of Applied Sciences
University of North Texas, Denton, Texas, USA

B.T. Walton, US Environmental Protection Agency
Research Triangle Park, North Carolina, USA

C.H. Ward, Department of Environmental Sciences and Engineering
Rice University, Houston, Texas, USA

Acknowledgments

We gratefully acknowledge financial support for the book, *Perchlorate Ecotoxicology*, through the Strategic Environmental Research and Development Program (SERDP) (CU-1235). We particularly appreciate the support of Mr. Bradley Smith, Dr. Jeff Marqusee, and Dr. Andrea Leeson in the SERDP office. The efficiency with which this textbook was developed at our Colorado meeting would not have been possible without the outstanding support of The Institute of Environmental and Human Health (TIEHH)/ Texas Tech University administrative staff, including Alicia Knight, Sam Hatchett, and Tammy Henricks. We appreciate the support of the Society of Environmental Toxicology and Chemistry in facilitating publication of *Perchlorate Ecotoxicology*. We are particularly appreciative of the excellent peer review comments of three anonymous reviewers. In addition, we appreciate the effort of Mimi Meredith in working closely with the co-editors to make the book publication as seamless as possible.

Contents

List of figures

List of tables

About the editors

Ronald J. Kendall is founding Director of The Institute of Environmental and Human Health (TIEHH), a joint venture between Texas Tech University and Texas Tech University Health Sciences Center at Lubbock, Texas. He is Professor and Chairman of the Department of Environmental Toxicology at Texas Tech University. Dr. Kendall has served as chair of the United States Environmental Protection Agency's Science Advisory Panel, as president of the Society of Environmental Toxicology and Chemistry, and on multiple panels of the National Research Council. He is Editor of Terrestrial Toxicology for *Environmental Toxicology and Chemistry*, has authored more than 200 refereed articles and books, and has won 136 research grants from federal, state, and foreign governments, industries, and foundations. He has received numerous awards, addressed the United Nations Committee on Sustainable Development, and consulted with many foreign countries on environmental issues.

Philip N. Smith, Ph.D. serves as Assistant Professor of Environmental Toxicology at The Institute of Environmental and Human Health, Texas Tech University. Dr. Smith is an ecotoxicologist whose research interests center on ecological and physiological characteristics of organisms, populations, and environments that contribute to contaminant exposure and adverse effects. His research examines pathways of contaminant exposure among mammals, birds, and a variety of aquatic organisms, trophic-level transfer of environmental contaminants, and physiological and population-level responses to contaminant exposure.

Authors

Todd A. Anderson, Ph.D.
Associate Professor
Texas Tech University

James A. Carr, Ph.D.
Associate Professor
Texas Tech University

Stephen B. Cox, Ph.D.
Assistant Professor
Texas Tech University

Kenneth R. Dixon, Ph.D.
Professor
Texas Tech University

W. Andrew Jackson, Ph.D., P.E.
Associate Professor
Texas Tech University

Ronald J. Kendall, Ph.D.
Professor/Director/Chair
Texas Tech University

Kevin P. Mayer
USEPA

F. M. Anne McNabb, Ph.D.
Professor
Virginia Tech

Reynaldo Patiño, Ph.D.
Professor
Texas Tech University

Ernest E. Smith, Ph.D.
Associate Professor
Texas Tech University

Philip N. Smith, Ph.D.
Assistant Professor
Texas Tech University

Erin M. Snyder, Ph.D.
Black & Veatch

Shane A. Snyder, Ph.D.
R&D Project Manager
Southern Nevada Water Authority

Christopher Theodorakis, Ph.D.
Assistant Professor
Texas Tech University

Jody Wireman, Ph.D.
Toxicologist
US Air Force

Contributing Authors

Eric Albers, Ph.D.
Texas Tech University

Jaclyn E. Cañas, Ph.D.
Texas Tech University

B. Angella Gentles, Ph.D.
Texas Tech University

Michael J. Hooper, Ph.D.
Texas Tech University

John Isanhart, M.S.
Texas Tech University

Mindy Landrum, M.S.
Environomics, Inc.

Scott McMurry, Ph.D.
Texas Tech University

Kui Tan, Ph.D.
Texas Tech University

Preface

The textbook *Perchlorate Ecotoxicology* is offered through SETAC Press as a contribution to establish the state-of-the-science of perchlorate in the environment and offer some perspective on the risk of this substance to ecological systems. Perchlorate contamination resulting to a large degree from Department of Defense operations, including development of rocket propellants, has resulted in widespread contamination, particularly in the western United States, and new scientific development has been needed to improve risk assessment on perchlorate. To evaluate the ecological risk perchlorate poses to military operation sites, the Strategic Environmental Research and Development Program, jointly funded through the Department of Defense, the United States Environmental Protection Agency, and the Department of Energy, provided funds to The Institute of Environmental and Human Health at Texas Tech University for implementation of a multi-disciplinary ecotoxicology research program to address perchlorate in the environment. Indeed, the book *Perchlorate Ecotoxicology* draws heavily from the funding received through the Strategic Environmental Research and Development Program and incorporates other leading scientists' research and their involvement with perchlorate research over the last few years.

Drawing upon research data development on perchlorate as well as a literature review, this book involved many months of planning and coordination through its co-authors, as well as a meeting 11-14 July, 2004, at Beaver Creek, Colorado, to bring all the chapters together for the book *Perchlorate Ecotoxicology*. To discuss *Perchlorate Ecotoxicology* as related to the risk assessment process drawing upon the United States Environmental Protection Agency's Risk Assessment Paradigm, the co-authors met and discussed these issues for several days, following months of planning. Although the book does not attempt to completely implement a full-scale ecological risk assessment, it does offer perspectives on data integration from multiple species in multiple studies, as well as drawing upon the literature review to give some perspectives from the co-authors into the ecotoxicological risk of perchlorate in the environment.

The main co-authors for *Perchlorate Ecotoxicology* are listed in the previous page, including name and affiliation. All of these persons participated fully in a collegial and multi-disciplinary perspective at the Beaver Creek meeting. Additional persons who offered input into individual chapters are noted as "Contributing Authors" to an individual chapter, even if they did not participate in the Beaver Creek meeting. All senior chapter authors did participate in the Beaver Creek meeting and fully support the conclusions reached by the group, particularly those related to the risk assessment-based chapter.

Perchlorate contamination continues to evolve as a national priority pollutant and as a substance of great debate regarding contamination of water supplies and the environment. We offer *Perchlorate Ecotoxicology* as a science-based text to improve our ability to implement technical-based risk assessment and to improve perchlorate evaluation and remediation goals and objectives in the future. The co-authors believe that the book will contribute to the developing science of addressing endocrine disruptors that oftentimes challenge environmental toxicologists by virtue of the potential for more subtle effects with perchlorate, which may require more complex experimental designs to evaluate. We embrace the definition of "endocrine disruptor" offered by the United States Environmental Protection Agency: "an environmental endocrine or hormone disruptor may be defined as an exogenous agent that interferes with the synthesis, secretion, transport, binding, action, or elimination of natural hormones in the body that are responsible for the maintenance of homeostasis, reproduction, development, and/or behavior" (Special Report on Environmental Endocrine Disruption: An Effects Assessment and Analysis, Risk Assessment Forum, United States Environmental Protection Agency, Washington, DC (EPA/630/R-96/012) February 1997). Based on this definition, we believe perchlorate is an endocrine disruptor.

The book is laid out in a chapter format to offer the following perspectives, including Chapter 1, "State of the Science: Background, History and Occurrence." This is followed by Chapter 2, "Environmental Fate of Perchlorate," which we think will offer an excellent update as to the degree of widespread perchlorate contamination in the environment and apparently not just for military-related activities. Chapter 3, "Thyroid Function and Mechanism of Perchlorate Action in Vertebrates," offers more perspective on modulation of endocrine systems by substances such as perchlorate. In Chapter 4, "Perchlorate Effects on Mammals," mammalian response in terms of thyroid endocrine toxicology is discussed in detail and related to environmentally relevant perchlorate exposures. In Chapter 5, "Perchlorate Effects in Birds," a discussion of perchlorate exposure to both adults and young birds is provided and the state of the science is addressed. In Chapter 6, "Effects of Perchlorate in Amphibians," the role of contaminants to amphibian decline is discussed as it relates to perchlorate ecotoxicology and sensitivity of amphibians to this substance. Chapter 7, "Perchlorate Effects in Fish," is related to perchlorate in the freshwater environment and aquatic toxicology consequences. This is followed by Chapter 8, "Perchlorate Effects on Invertebrates," in which we review the state of the science on the response of invertebrates to perchlorate, particularly as they contribute to the food chain. In Chapter 9, "Application of Population Models to Perchlorate Ecotoxicology," we address population biology perspectives as they relate to perchlorate exposure and effects discussed in previous chapters. This will offer the reader some perspective from the co-authors on population effects from perchlorate exposures. In the final chapter, "Ecological Risk Assessment," we integrate the information developed in the previous chapters to offer an application of the ecological risk assessment paradigm toward a better understanding of perchlorate ecotoxicology.

We appreciate the opportunity to work with SETAC Press in publication of *Perchlorate Ecotoxicology* and its contribution to a better understanding of the science of endocrine disruptor research and its relationship to ecotoxicology. This book follows a previous SETAC Press book *Principles and Processes for Evaluating Endocrine Disruption in Wildlife*, edited by Kendall et. al., which contributed early on in the endocrine disruptor science debate related to how we should be implementing science-based research to more fully evaluate endocrine disruptors and their relevance to ecotoxicological consequences. We thank SETAC for offering the opportunity to continue to contribute to this complex area of science, and we appreciate the funding provided for research and our meeting through the Strategic Environmental Research and Development Program.

Sincerely,
Ronald J. Kendall, Ph.D. and Philip N. Smith, Ph.D.
Co-Editors

one

State of the Science: Background, History, and Occurrence

Kevin P Mayer, W Andrew Jackson, Shane A Snyder, Philip N Smith, Todd A Anderson

Introduction

Emerging environmental concern

Perchlorate, a chemical that had been known both in nature and through synthesis since the 19th century, only emerged into the limelight as an environmental contaminant in 1997. No formal state or federal regulatory level, such as a drinking water standard, has been established for this chemical as of 2005, other than several site-specific waste cleanup decisions. Local and nationwide attention grew rapidly over the last several years as improved analytical methods led to surprisingly widespread detections of perchlorate in the environment. Although perchlorate (ClO_4^-) has several important uses in its various salt forms and as perchloric acid, this simple inorganic chemical had been relatively unknown outside of a few specific research and industrial settings. By 2004, perchlorate issues in the United States had been featured prominently in national newspapers and magazines, in state and national legislative actions, and in numerous national research conferences, and even received attention from the White House in a decision to seek advice from the National Academy of Sciences (NAS). Much of the interest and controversy surrounds estimates of risks to human health. The NAS report (National Research Council [NRC] 2005) recommended an oral reference dose that was adopted by the United States Environmental Protection Agency (USEPA) in February 2005 as a total level of perchlorate that would be protective of human health. Neither the NAS nor the USEPA has resolved important issues such as quantifying the amount of perchlorate

Perchlorate Ecotoxicology. Ronald J Kendall and Philip N Smith, editors.

ISBN-10: 1-880611-90-2 ISBN-13: 978-1-880611-90-6

in food (e.g., see Kirk et al. 2005). Other scientific and technical disciplines besides toxicology play important roles in addressing perchlorate in the environment, including analytical chemistry, groundwater and surface water hydrology, soil science, water treatment engineering, agricultural sciences including fertilizer, and the topics featured in this book, ecotoxicology. This chapter provides a brief background and historical perspective on perchlorate and an examination of the increasing body of information on the environmental occurrence of perchlorate in the US. Possible ecosystem exposure related to occurrence of perchlorate at specific sites and groundwater sampling locations is also discussed.

Basic chemical properties

Forming salts with different cations

The environmental concern over perchlorate in soil or water focuses on the perchlorate anion, which originates from the dissolution of perchloric acid or perchlorate salts formed with one of a number of cations. The most common manufactured perchlorate salts are ammonium, potassium, sodium, and magnesium perchlorates, although a long list other perchlorate salts can form with other metals and with inorganic and organic cations (Schilt 2003) (Table 1-1). The common perchlorate salts are considered extremely soluble in water, with the exception of sparingly soluble potassium perchlorate. At 25 °C, about 25 g of ammonium perchlorate will dissolve in 100 mL of water, while only 2 g of potassium perchlorate will dissolve in the same amount of water. In units that are commonly used in regulatory parlance, these solubilities correspond to 250 million and 20 million $\mu g \cdot L^{-1}$ (parts per billion) respectively. Once perchlorate salts have dissolved in water, the cation is rarely environmentally relevant unless the cation itself has an effect, such as the toxicity of ammonium. There are numerous examples of ammonium perchlorate being transported into groundwater at considerable perchlorate concentrations with little or no ammonium detectable (Batista et al. 2003).

Thermodynamics and energetics

In the solid state, x-ray diffraction shows that perchlorate forms a nearly perfect tetrahedron, with the single chlorine atom surrounded by the 4 oxygen atoms. Likewise, the geometry of perchlorate in aqueous solution is likely to be perfectly tetrahedral and symmetrical in 3 dimensions (Urbansky 1998). The single anionic charge is highly delocalized, appearing over the entire surface of the molecule. Perchlorate has a molecular volume of 44.5 $cm^3 \cdot mol^{-1}$ at 25 °C, which is similar to that of iodide

Table 1-1 Physical properties of ammonium and alkali metal perchlorates at 25 °C

Perchlorate cation	NH_4	Li	Na	K	Rb	Cs
Density	1.95	2.43	2.50	2.53	2.9	3.33
Solubility in water (g per 100 g water)	25	60	210	2.1	1.3	2.0
Heat of formation ($kcal \cdot mol^{-1}$)	–69	–92	–92	–104	–104	–104

Data from Schilt 2003

($36.7\ cm^3 \cdot mol^{-1}$). The relatively low charge density reduces the affinity of perchlorate anions for cations, leading to high water solubility. This geometry also plays a role in the energetics of perchlorate.

The valence state of the chlorine atom in perchlorate is +7, indicating that perchlorate is strongly electrophilic and acts as a potent oxidizer. The Gibbs free energy of reaction for ammonium perchlorate has been calculated at –1268 kJ, indicating that the decomposition of ammonium perchlorate releases an extremely large quantity of energy. When 4 moles (468 grams, or slightly more than a pound) of ammonium perchlorate burn, enough energy is released to lift a 1 kg object 130 km (Urbansky 1998). Although the energy of reaction might suggest that this reaction could be spontaneous, free energy is not released at room temperature because the reaction kinetics are too slow for the reaction to occur. The activation energy needed to alter the stable tetrahedral barrier formed by the 4 oxygen atoms can be induced by heat of about 300 °C (Schilt 2003). Such stability, combined with the very high release of energy when the reaction is intentionally initiated, makes perchlorate salts highly valued as propellants and explosives.

Stability and mobility in environment

The kinetic stability of the perchlorate anion is also exhibited in the resistance of perchlorate to react under typical environmental conditions. Abundant reducing forces are often present in groundwater aquifers, yet this strong oxidizer remains present. Several perchlorate plumes have been studied in the western US where groundwater contamination likely occurred decades before. The capabilities and limitations of biological perchlorate degradation will be discussed in Chapter 2.

The low charge density of perchlorate also is responsible for its minimal reaction with mineral and soil surfaces. Because the perchlorate anion does not impact cationic sorption and desorption phenomena (chemical attraction to charged surfaces), perchlorate is extensively used as a counter anion in studies of surface chemistry of metal cations in solution. The structure and chemical properties of perchlorate create the potential for a persistent, rapidly spreading environmental contaminant.

History

Historical interest in environmental perchlorate

Several monographs provide a thorough historical perspective on perchlorate (Schumacher 1960; Urbansky 2002; Schilt 2003). Some notable milestones include the synthesis and identification of perchlorate early in the 19th century. By the 1880s, perchlorate was identified by European researchers as an unwanted salt in sodium nitrate fertilizer derived from unusual salt deposits in the Atacama Desert in Chile. Perchlorate was undesirable because it was implicated in reducing crop productivity when the concentration was too high. The energetic properties of perchlorate raised the interest in perchlorate as a component of propellants and explosives, which stimulated considerable research on the chemistry and production of perchlorate in the 19th and early 20th centuries. World War II and especially the Cold War drove a rapid expansion in perchlorate production in the US for use in propulsion and related applications.

Medical and biological effects: perchlorate and iodide

Perchlorate competes with iodide for uptake into the thyroid, and this mechanism of blocking an essential component of thyroid hormones is still considered the primary human health and ecological concern. Perchlorate salts have been used pharmaceutically to treat overactive thyroids, primarily in the 1950s and 1960s.

As a compound that has an effect on the thyroid, perchlorate can be classified as an endocrine disruptor. According to the USEPA, "an environmental endocrine or hormone disruptor may be defined as an exogenous agent that interferes with the synthesis, secretion, transport, binding, action, or elimination of natural hormones in the body that are responsible for the maintenance of homeostasis, reproduction, development, and/or behavior" (USEPA 1997). Furthermore, the Endocrine Disruptor Screening and Testing Advisory Committee (EDSTAC) specifically named the thyroidal hormone system as one of the endocrine systems that should be considered in USEPA's Endocrine Disruptor Screening Program (EDSTAC 1998).

The availability of iodide in the diet of humans and other animals must be considered as a potentially important factor in the ecotoxicity of perchlorate. In many parts of the US, the availability of environmental iodide may be sufficient to confound the toxicological effects of perchlorate. Conversely, iodide deficiency in the soil and water of some geographical regions is well documented (Hetzel 1989).

Early analytical methods

The low charge density and minimal reactivity of perchlorate posed some challenges for perchlorate analytical methods until early in 1997, when the State of California published a method that used ion chromatography. The lack of reliable analytical tools can be best exemplified by a 1985/1986 effort by the USEPA to examine perchlorate in an aquifer near a Southern California rocket manufacturing plant. Perchlorate was detected in samples of 14 wells ranging from 110 to 2600 $\mu g \cdot L^{-1}$ by a contract laboratory using a proprietary colorimetric method. The Regional USEPA office quickly requested an evaluation of potential health effects from the Centers for Disease Control (Takata 1985; Margolis 1986). However, the data validation process indicated that the results could not be confirmed, and in the absence of a valid analytical method for perchlorate, the immediate focus shifted to other contaminants present such as the chlorinated solvents. When this site was resampled in 1997 and subsequent years, a large plume of perchlorate was confirmed at concentrations exceeding 100 $\mu g \cdot L^{-1}$ in some wells.

Natural formation and occurrence

Schumacher (1960), Schilt (2003), and others have commented on the occurrence of perchlorate in natural salt deposits in the Atacama Desert of Chile. How the perchlorate was produced and why it appeared only in that location was described as a "disturbingly unanswered" question. Erickson (1983) provided some detailed insight on the unique mineralogy of the Atacama, including perchlorate and iodate salts found there. He suggested that the remarkably arid conditions in this desert have been continuous over millions of years. These conditions could lead to a slow accumulation that would otherwise be washed away and possibly degraded microbially. Erickson speculated that the oxidation of chlorine to perchlorate could occur in the upper atmosphere where high-energy radiation and the presence of ozone could result in some perchlorate production. Recent analysis of chlorine and oxygen stabile isotope ratios provide some evidence for this mechanism and may enable investigators to differentiate between manmade and naturally formed perchlorate (Bao and Gu

2004). Another possible mechanism of atmospheric generation includes electrochemical production from lightning. Alternatively, it is possible that oxidation of chloride at mineral surfaces through evaporative weathering may be responsible for the natural occurrence. Orris et al. (2003) searched evaporites in the western US for perchlorate and discovered trace levels in a number of locations. The overall contribution of natural deposits as a significant source of perchlorate in usable water resources is relatively unknown outside of Chile. Some preliminary information using state-of-the-art analytical methods indicates that perchlorate may be detected in groundwater and surface water at concentrations below 1 $\mu g \cdot L^{-1}$ in areas not associated with manmade perchlorate use or disposal. For instance, the Southern Nevada Water Authority (SNWA) has reported trace concentrations of perchlorate in the upper Colorado River using mass spectrometric detection (SNWA 2004). Snyder et al. (2005) showed trace levels of perchlorate in a variety of natural and bottled waters, indicating a naturally occurring source is likely. Another possible example that has been relatively well studied is the occurrence of perchlorate in groundwater in the vast High Plains region of western Texas (Jackson et al. 2004).

Fertilizer

Urbansky et al. (2001) conducted rigorous analytical and nationwide sampling efforts to examine the potential for perchlorate in fertilizers. Several earlier studies of fertilizers reported that perchlorate might be present in additional fertilizer products. Urbansky found measurable levels of perchlorate only in fertilizer derived from Chilean nitrate deposits. He reported that processing methods for this product have been changed to minimize perchlorate content that was sometimes as high as 1% or more. However, more sensitive analytical techniques capable of trace-level detection may be able to discover perchlorate in sources that were previously undetectable (Snyder et al. 2005).

Uses of perchlorate

Perchlorate salts have been used in a diversity of industrial and research applications. However, in 1998 USEPA asked the 2 existing perchlorate manufacturers to describe the uses of perchlorate. Nearly all, possibly more than 90%, of the ammonium perchlorate manufactured in the US is for use in aerospace and other propulsion systems. The solid rocket fuel boosters of the space shuttle are fueled by a mixture of approximately 15% metal powder and 70% ammonium perchlorate, combined with a form of rubber as a binder for the remaining 15%. Each space shuttle launch uses 2 solid rocket fuel boosters that consume more than 1.5 million pounds of perchlorate (mass of the anion alone). The rubber binder has a limited useful life of approximately 10 to 20 years, and the solid rocket fuel must be washed out of the rocket and replaced on a routine schedule. High-pressure water has been used for cleaning out the old fuel, creating considerable wastewater that often was disposed to the environment. Such operations, along with testing and disposal of rockets, led to some of the first discoveries of potential environmental issues.

Pyrotechnics and explosives comprise most of the other applications for perchlorate. Perchlorate imported from places such as China could be used when less stringent specifications are necessary. These uses might include flares, pyrotechnics, and smokeless powder for muzzle-loading guns. Other formerly common uses, such as metal plating, apparently are no longer significant.

Manufacture

Oldbury Electro Chemical Company in Niagara Falls, New York, manufactured perchlorate as early as 1909. Production expanded to other locations and companies during and after World War II. Millions of pounds of perchlorate were manufactured each year, in some years exceeding 20 to 30 million pounds (USEPA 1998). Although 9 different plants have operated at various times throughout the years, only 2 manufacturing plants were in operation after 1975. Since the late 1990s, there has been only a single manufacturing facility in the United States.

History of Environmental Interest

California post–World War II

Some documents and letters from the period around World War II express initial concerns by water quality officials from the State of California that rocket manufacturing activities were flushing salty water into important groundwater aquifers. California's arid climate and expanding population place intense demand on all potential water resources and have resulted in stringent water quality protection regulations. Lack of analytical methods and limited information on toxicity prevented much follow-up on these early perchlorate disposal issues.

Cleanup of groundwater sites for other contaminants

San Gabriel Valley

With the establishment of state and federal toxic waste cleanup legislation in the early 1980s, additional resources and enforcement regulations precipitated more investigations into toxic chemicals in the environment. The San Gabriel Valley Superfund Sites in southern California became federal cleanup sites largely due to extensive areas of the aquifer being contaminated with chlorinated solvents. USEPA asked one of the large users of the solvents about other chemicals that could have been released to the environment. This user indicated that perchlorate contamination was possible, and efforts began to monitor the groundwater for perchlorate and other compounds. However, the monitoring results were eventually considered inconclusive. Advice from the Centers for Disease Control on the toxicity of perchlorate was also inconclusive.

Rancho Cordova

Another large rocket manufacturing facility in northern California also became a Superfund site in the 1980s and faced a similar set of potential groundwater contaminants. Monitoring conducted around 1990 discovered perchlorate concentrations above 1000 $\mu g \cdot L^{-1}$ in an aquifer that was used as a drinking water supply. Analytical methods were available that could confirm the presence of perchlorate at this concentration. Superfund cleanups are based both on promulgated regulatory levels and on a risk assessment for the site. The regional staff requested guidance from the toxicologists at USEPA's National Center for Environmental Assessment. In 1992, a provisional estimate based on the little research available suggested that a perchlorate dose equivalent to approximately 4 $\mu g \cdot L^{-1}$ could be protective of human health. There were major gaps in important scientific information, for which the protocol required application of uncertainty factors. It should be noted that until 1997, the analytical methods available were reliable only to about 400 $\mu g \cdot L^{-1}$.

Regulatory process: dealing with uncertainty

Establishing a formal regulation is a lengthy process involving many scientific disciplines as well as economic and other stakeholder issues. The first step is a scientific analysis of available research pertinent to the contaminant. The US Department of Defense (DoD) and the aerospace industry expressed concern about USEPA's 1992 and 1995 provisional assessments of perchlorate toxicity. In 1997, they met with USEPA and agreed to fund a series of studies designed to address the missing information. In 1998,

USEPA conducted a review of the partial information from these studies, and in 2002, USEPA published an external review draft toxicity assessment. The 2002 draft suggested that approximately 1 $\mu g \cdot L^{-1}$ might be a protective level for the most sensitive populations (such as pregnant women and newborns). The National Research Council (NRC 2005), recommended 0.0007 $mg \cdot kg^{-1} \cdot d^{-1}$ as an oral reference dose, an estimate of the total level of perchlorate that would be protective of human health. The NRC recommendation was adopted by USEPA in February 2005 and published on the USEPA's Integrated Risk Information System database. Neither the NRC nor the USEPA have resolved important issues such as quantifying the amount of perchlorate in food, and a decision on the need for a federal standard for perchlorate awaits further information gathering.

For site-specific cleanups, USEPA and individual states have been making site-specific decisions when a cleanup project cannot wait for a formal standard. USEPA is currently using a range of values from 4 $\mu g \cdot L^{-1}$ to 18 $\mu g \cdot L^{-1}$. Most of the formal decisions thus far have recommended the lower end of the range. At one USEPA cleanup project, bottled water has been supplied to community members whose wells exceed 18 $\mu g \cdot L^{-1}$ of perchlorate, unless specific health considerations warrant response at a lower concentration threshold. At several Superfund sites, USEPA selected 4 $\mu g \cdot L^{-1}$ as a treatment objective when the groundwater contains hundreds or thousands of $\mu g \cdot L^{-1}$ perchlorate.

California, New Jersey, and Massachusetts have begun the process of developing state standards for perchlorate. In 2006, Massachusetts adopted 2 $\mu g \cdot L^{-1}$ as a drinking water standard for their state. In 2004, California established a Public Health Goal for perchlorate in drinking water at 6 $\mu g \cdot L^{-1}$, and proposed this level as a drinking water standard in 2006. New Jersey has proposed 5 $\mu g \cdot L^{-1}$ as a protective standard. Six other states have various advisory levels or monitoring requirements.

Six other states have various advisory levels or monitoring requirements. Maryland and New Mexico use 1 $\mu g \cdot L^{-1}$ as advisory levels or monitoring levels respectively, Texas recommends 4 $\mu g \cdot L^{-1}$ as a drinking water advisory level for sensitive populations and 17 $\mu g \cdot L^{-1}$ as a residential cleanup levels, New York State has a 2-tier action level of 5 and 18 $\mu g \cdot L^{-1}$, Arizona established a Health-Based Guidance Level of 14 $\mu g \cdot L^{-1}$, and Nevada is reported to use 18 $\mu g \cdot L^{-1}$ as a drinking water advisory level. None of these are enforceable standards.

Occurrence in the Environment

Until recently, there has not been a coordinated nationwide effort to monitor for perchlorate in the environment. It is important to note that there are a variety of objectives for monitoring even a single chemical and that the results must be understood in the context of the objectives and protocols. A list of known perchlorate detections in the US, current as of August 2004, has been compiled at an USEPA website: http://www.epa.gov/fedfac/documents/perchlorate.htm.

Including information from the August 2004 public release of USEPA's Unregulated Contaminant Monitoring Rule (UCMR) preliminary results and the DoD reports, perchlorate has been detected in the water in 35 states and in Puerto Rico.

Two additional factors of the existing occurrence information must be emphasized, particularly in the context of ecotoxicology. One is that analytical capabilities largely limit reporting of perchlorate to concentrations above 4 $\mu g \cdot L^{-1}$. The distribution pattern for both groundwater and drinking water could be substantially expanded if the analytical limit decreased significantly. It is conceivable that there may be background concentrations in the nanogram per liter range, particularly in water from arid regions.

A second caveat of the occurrence data is that the majority of the monitoring efforts focus on human exposure through drinking water pathways, often through groundwater. Groundwater that does not impact surface water or soil within the root zone may not have an ecological effect. However, each of the groundwater sources has the potential to affect a larger area depending on its use and discharge as well as on its indirect paths, such as transport into plants. Other than specific sites (e.g., some of the industrial and DoD sources discussed later), there have been few attempts to directly measure interactions between groundwater and ecosystems such as surface water, soil, and plant tissue.

Information from site investigations

Much of the early information on perchlorate occurrence in the environment came from investigations of specific facilities, such as those at federal or state groundwater cleanup projects. A few examples include the San Gabriel Valley Superfund Site, Olin San Martin, and the Lockheed Redlands state projects in California; the Basic Magnesium Inc. (BMI) state-led project in Nevada; Longhorn Army Ammunition Depot in Texas; and the Massachusetts Military Reservation. In such cases, a substantial set of information about the site and groundwater movement was already available, and monitoring wells were in place. These detailed investigations often provide the best information on the origin of the perchlorate because identification of the contaminant source is one of the objectives of the monitoring.

Although groundwater usually is the focus of the site investigations, soil and surface water contamination relevant to potential ecosystem exposures have been measured at a number of sites (Parsons et al. 2001; Smith et al. 2001). Streams, ponds, and soil contamination areas at Longhorn Army Ammunition Depot, for example, have been extensively studied for ecosystem effects.

A former Kerr-McGee perchlorate manufacturing facility near Las Vegas, Nevada, is another important example. This site was recognized in 1997 as the source of nearly 1000 pounds of perchlorate entering Lake Mead and the Colorado River each day. Efforts over the last few years have reduced this to about 100 pounds per day, and the levels are expected to continue to decrease (Figure 1-1). Perchlorate remains detectable through-

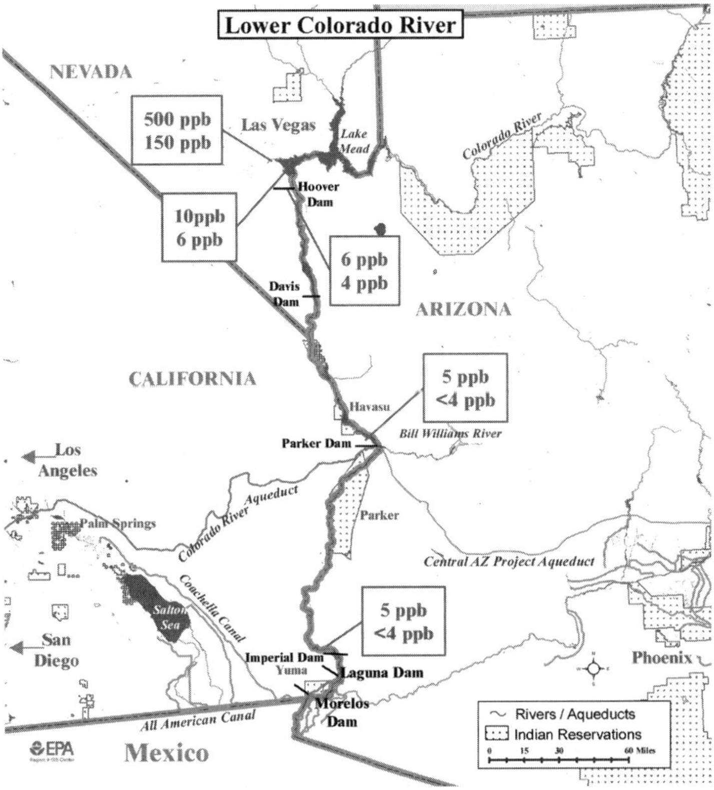

Figure 1-1 Perchlorate concentrations in the Lower Colorado River System in 2002, followed by 2004

out more than 300 miles of the Lower Colorado River, downstream of Lake Mead and the Hoover Dam. The Colorado River is an essential source of municipal water for 3 states and is the primary source of irrigation water for important agricultural areas in this part of Arizona and California.

At Kerr-McGee, Longhorn, and many other sites, active cleanup responses are addressing perchlorate as a contaminant of concern. Actual concentrations of perchlorate in various environmental locations will change as cleanup efforts become increasingly effective.

The decision to search for perchlorate at a particular cleanup site is often related to knowledge regarding perchlorate use at the facility. It is not unexpected that most of the confirmed releases occur at aerospace or propulsion manufacturing, testing, or disposal facilities and in pyrotechnic, flare, or explosives operations.

Unregulated Contaminant Monitoring Rule (UCMR)

Since 2001, USEPA's Office of Water has included perchlorate in a nationwide program to collect data for potential drinking water contaminants for which no federal standard has been established (maximum contaminant level, or MCL). The program objectives include determinations whether a national drinking water standard may be warranted and whether the actual occurrence in regulated public drinking water supply systems is widespread or localized. As of August 2004, the database included results from 3460 Public Water Supply Systems nationwide. Just over 4% (145 systems) reported perchlorate detections above 4 $\mu g \cdot L^{-1}$. Eventually, about 3800 systems are expected to provide data.

These data rarely provide clear information on the sources of the contamination, although in some cases positive results motivated a search for possible sources. The UCMR database identifies whether the water supply is derived from groundwater or surface water. Absent more site-specific knowledge, UCMR data cannot readily be used to quantify exposure to ecosystems.

State, local, and regional monitoring

Certain state and local officials have undertaken area-wide monitoring efforts. Area-wide monitoring efforts typically are spurred by state and local public health officials who wish to inform themselves and their communities of the potential for perchlorate contamination. The greater density of sampling points is more conducive to deriving information on possible sources and potential ecosystem exposure.

State and local monitoring efforts are often directly or indirectly associated with site-specific investigations and the federal UCMR as well. Of the 583 UCMR samples with perchlorate detections in 26 states plus Puerto Rico, 368 detections were in California and 48 were from Long Island, New York. These 2 areas have some of the most thorough area-wide perchlorate monitoring in the US, along with Massachusetts, southern Nevada, and parts of Texas. Since 1998, USEPA has noted that the number of detections seem to be related to the effort made to search for perchlorate (USEPA 1998).

California has an extensive monitoring program that has tested about 6500 public water supply wells throughout the state. Of these, there are approximately 350 drinking water supply wells with perchlorate at or above 4 $\mu g \cdot L^{-1}$ and 250 at or above the recent action level of 6 $\mu g \cdot L^{-1}$. Suffolk County, New York, included perchlorate in an extensive routine well monitoring program, with some investigative follow-up when a pattern of perchlorate was noted. Massachusetts tested 600 public water supply systems and reported 10 systems with detections at or above 1 $\mu g \cdot L^{-1}$. Southern Nevada officials analyze a prodigious number of water samples to monitor for perchlorate through Las Vegas Wash and Lake Mead. New Mexico officials began an initiative to check for perchlorate in wells near installations that may have used perchlorate, such as munitions storage and disposal facilities, and they have expanded the effort to examine groundwater in several agricultural communities with nitrate fertilizer use. Oregon, Utah, and Arizona have also begun perchlorate monitoring programs, and it is likely that other states will follow.

Results of such area-wide monitoring often can identify potential sources, such as the groundwater plume traced to a fireworks disposal area in New York. Just as often, the results identify the need for a more focused investigation. In California, state and federal programs are using the results from the statewide drinking water source monitoring to prioritize regulatory investigations.

A USEPA regional office in the Midwest worked with state officials in Nebraska and Iowa to check water quality in rural farm communities, and they included perchlorate as an analyte to examine whether fertilizer use could be related to perchlorate contamination. One shallow well in Nebraska with a confirmed level of 5 $\mu g \cdot L^{-1}$ was in the vicinity of a facility that had blended specialty fertilizers. Another detection in Iowa was near a fertilizer wholesale operation, but further investigation traced the source to an area that seemed to preclude the fertilizer facility. Perchlorate detections in some agricultural locations in California, New Mexico, and

New York have not eliminated the potential of a fertilizer link. Further investigations are needed before conclusions can be drawn.

Regional water quality officials have suggested that at least one unexplained cluster of perchlorate-contaminated wells in Riverside County, California, might have originated from prolonged application of Chilean nitrate fertilizer at a demonstration orange grove around the time of World War II. This site also is the subject of further investigations.

Texas High Plains region

Groundwater from the High Plains region of Texas and eastern New Mexico has been extensively studied following the detection of perchlorate in samples collected under the UCMR program. The study area included more than 135 000 km^2 and a number of geologic formations and aquifers. Public water system wells, private wells, US Geological Survey (USGS) monitoring wells, and local water district wells were surveyed. Perchlorate was found to be present over the entire study area at concentrations ranging from 0.1 $\mu g \cdot L^{-1}$ to greater than 50 $\mu g \cdot L^{-1}$. Numerous potential sources for this perchlorate were carefully examined. Currently, only 2 potential sources remain likely: historic use of Chilean fertilizer and/or perchlorate as a natural constituent in the regional geology. While no definitive method was available to differentiate these potential sources, a variety of indirect data sources and studies, including land use studies, transport modeling, mass calculations, and historical fertilizer use records, indicate that the most likely source is natural (Jackson et al. 2004).

Groundwater in this region is generally more than 15 m below ground and can be more than 120 m below ground. However, significant surface exposures exist because of large amounts of irrigation. Perchlorate has been found in surface soil as well as in plants grown in the area.

Department of Defense information

The DoD has provided some perchlorate monitoring data from military facilities that are included in the list compiled by USEPA. A range of operations has been associated with perchlorate releases at military facilities including open burn–open detonation, research and testing, and some training activities. A systematic and comprehensive monitoring program for current and former military sites has not been undertaken yet. Some of the facilities have extremely thorough investigations with full data quality control, while others have screening-level results that may or may not be confirmed. As more data become available, a more complete picture of perchlorate occurrence is gradually being developed.

Ecosystem Exposure

Source and groundwater relevance to ecosystems

With the information currently available, it is not feasible to estimate ecological exposure at most of the locations where perchlorate has been detected. Quantitative data for the majority of the known environmental occurrence locations are based on groundwater measurements. For example, the list compiled by USEPA in 2003 notes soil or surface water contamination for 16 of the 95 sites (Mayer 2003). Simple lists and concentrations shed little light on the magnitude of environmental effects, whether relatively insignificant or affecting a wide region. Some of the instances where perchlorate is known in surface water and soil can generate ecological exposures far beyond the site property boundary, such as the Kerr-McGee release to the Las Vegas Wash in southern Nevada. Absence of direct sampling of surface water or soil certainly does not preclude the potential for exposure to plants and to aquatic and terrestrial fauna. It is highly unusual for any contaminant to be released directly to groundwater without passing through soil or surface water. Additional investigations of these sites may provide more complete data on perchlorate exposure to various ecosystem components.

Groundwater occurrence itself has the potential to become an ecological exposure. Particularly in arid regions, groundwater is widely used for irrigation, bringing subsurface chemicals to surface soils and to surface water that drains the irrigated lands. Contaminated groundwater may also resurface in seeps, springs, wetlands, ponds, and streams.

Measurements of perchlorate exposure in ecosystems

Several sites with known perchlorate releases to the environment were studied in greater detail to provide ecological exposure data. These sites are useful for preliminary estimates for perchlorate exposure values and are instructive of the potential for exposure from other sites that have not been examined for ecological exposure.

Smith et al. (2001) provided a preliminary assessment of perchlorate concentrations in environmental compartments and tissues of various animals at the Longhorn Army Ammunition Plant (LHAAP), Karnack, Texas. The LHAAP is composed of a variety of habitats, including mixed upland hardwood and pine forest, interspersed with a number of permanent and ephemeral drains, most of which eventually flow into Caddo Lake on the east side of the site. Some bottomlands with mixes of hardwoods and bald cypress act as drains. The LHAAP contains areas contaminated with perchlorate: 2 burning grounds, a water-treatment holding pond, and a building used for perchlorate handling. Military-related activities at the LHAAP, primarily manufacture of solid propellant rocket motors and demilitarization of rockets, distributed ammonium perchlorate across portions of the facility and into groundwater beneath the site.

Perchlorate was detected in water and/or sediment and in soil samples from all the sites sampled at LHAAP. With the exception of one highly contaminated water treatment holding pond, perchlorate concentrations in all other samples were less than 85 ppb. Perchlorate concentrations in the pond ranged from 30 000 to 31 000 $\mu g \cdot L^{-1}$ in water and from 12 000 to 35 000 $\mu g \cdot L^{-1}$ in sediment pore water.

Vegetation contained the highest levels of perchlorate of all samples analyzed. Bulrushes (*Scirpus* sp.) growing out of the highly contaminated pond contained perchlorate at concentrations of 7620 ± 1460, 4450 ± 2240, and 840 ± 410 $ng \cdot g^{-1}$ dw in above-waterline tissues, below-waterline tissues, and roots, respectively. Crabgrass (*Digitaria* sp.), cupgrass (*Erichloa* sp.), and goldenrod (*Solidago* sp.) samples collected from areas adjacent to a perchlorate grinding facility (Building 25C) contained extremely high concentrations of perchlorate (1060, 1030, and 5557 $\mu g \cdot g^{-1}$ dw) in blades or leaves. Perchlorate concentrations in the leaves of goldenrod were higher than in its seeds, stems, and roots. Perchlorate concentrations in seeds from both goldenrod and crabgrass were 184 and 1880 $\mu g \cdot g^{-1}$ dw, respectively.

Mammalian wildlife becomes exposed to perchlorate through ingestion of contaminated water, soil, sediment, vegetation, and aquatic vertebrates and invertebrates. At LHAAP, perchlorate was found in tissues of several species of fish, amphibians, and aquatic insect larvae. Highly water-soluble substances like perchlorate generally do not bioconcentrate, and perchlorate bioconcentration was not evident in samples acquired from the LHAAP. For example, bullfrog tadpoles and damselfly larvae collected from the highly contaminated pond had far lower concentrations of perchlorate than was present in water, although the body burdens (ranges from 811 to 2567 $ng \cdot g^{-1}$) were comparable between the 2 taxa. Concentrations in largemouth bass, sunfish, weed shiners, mosquitofish, and blackstripe top minnows ranged from below the detection limit to 206 ng perchlorate$\cdot g^{-1}$. Concentrations of perchlorate in frogs and toads ranged from below the detection limit to 580 $ng \cdot g^{-1}$. Based on these findings, it was evident that mammalian wildlife inhabiting the aquatic–terrestrial interface could become exposed to perchlorate through ingestion of contaminated water and food.

Parsons et al. (2001) reported similar findings of perchlorate contamination at 6 sites. These include

1) sites associated with irrigation water from the Colorado River in the vicinity of Yuma, Arizona;
2) Las Vegas Wash and Lake Mead near Las Vegas, Nevada;
3) Allegheny Ballistics Laboratory, Rocket Center, West Virginia;
4) Holloman Air Force Base, Otero County, New Mexico;
5) Naval Surface Warfare Center, Indian Head, Maryland; and
6) Longhorn Army Ammunition Plant, Karnack, Texas.

Except for those in the vicinity of Yuma, all of these sites are associated with localized contamination related to the manufacture, handling, or use of perchlorate in solid propellants. The Yuma sites are approximately 400 km downstream along the Colorado River from the Las Vegas Wash and Lake Mead sites. Parsons et al. (2001) suggest that there is no localized source of the perchlorate; therefore, the most likely source of perchlorate contamination in these soils is believed to be Colorado River irrigation water (USEPA 1998).

A spring associated with the Las Vegas Wash site had concentrations of 1.0 $\mu g \cdot L^{-1}$ to 130 $mg \cdot L^{-1}$ in surface water, while concentrations in the highly contaminated pond pond from the LHAAP site ranged from 3500 to 3800 $\mu g \cdot L^{-1}$ in September 2000. In the vicinity of a sediment delta created by the Las Vegas Wash in Las Vegas Bay of Lake Mead, Parsons documents a maximum perchlorate concentration of 68 $\mu g \cdot L^{-1}$ in surface water. The field studies by Parsons et al. (2001) found that, for various sites, wet-weight perchlorate concentrations in terrestrial vegetation samples were 1.5× to 80× the wet-weight concentrations in soil samples. The maximum measured concentration in vegetation at irrigated sites in the vicinity of Yuma, Arizona, was 1.0 $mg \cdot kg^{-1}$ wet weight. At sites with soil contamination related to the manufacture, handling, or use of perchlorate in solid propellants, maximum plant concentrations were

- 428 $mg \cdot kg^{-1}$ wet weight at a spring;
- 99.2 $mg \cdot kg^{-1}$ wet weight at a site upstream from Lake Las Vegas in the Las Vegas Wash area of the Lake Mead Recreational Area, Nevada; and
- 300 $mg \cdot kg^{-1}$ wet weight at the Burn Area of Allegheny Ballistics Laboratory, West Virginia.

Detection limits were reported as approximately 0.4 $mg \cdot kg^{-1}$ wet weight for most vegetation samples.

Conclusions and Data Needs

The data for perchlorate occurrence in the US has largely focused on protection of drinking water resources, with relatively few investigations considering ecological exposures. A substantial set of occurrence data, such as the national monitoring program under the UCMR, provides very limited information on the actual sources of perchlorate, much less on ecological fate. Where adequate investigations have been conducted, particularly at sites with relatively large amounts or high concentrations of perchlorate, a limited number of aerospace, military, pyrotechnical, explosive, and disposal operations have been implicated. Contributions from nonpoint sources, such as traces in certain fertilizer or natural formation, are still under study at some of the sites.

Some important studies have been conducted at specific locations to expand basic information on perchlorate detection to quantitative measurements of ecosystem fate and exposure. There may be more useful information on perchlorate ecological exposure and fate in site investigation reports that are not readily available to the scientific community. Clearly, there is a pressing need for ecological exposure information at specific sites and for expanding the information base available for general use in perchlorate ecotoxicology.

Most of the occurrence data is limited to concentrations greater than 4 $\mu g \cdot L^{-1}$. This limit is adequate for investigations at sites where relatively large amounts of perchlorate were manufactured, handled, or disposed. Use of more robust analytical methods to examine distribution at levels less than 4 $\mu g \cdot L^{-1}$ is already beginning to fill out the picture of ecological exposure to perchlorate, including the potential for a background concentration in some regions in the nanogram per liter range. Future research on ecotoxicology would benefit greatly by a review of existing data and standardized analytical methods with lower detection limits.

Basic research is also needed on the natural formation and accumulation of perchlorate. Some effort has begun to develop stable isotope fingerprinting methods to attempt to differentiate between manmade and naturally formed perchlorate. There are legitimate questions on the conditions under which perchlorate formed in nature may accumulate in the environment and whether this source of perchlorate may constitute a significant ecological exposure.

References

Bao H, Gu B. 2004. Natural perchlorate has a unique oxygen isotope signature. Environ Sci Technol 38:5073–5077.

Batista JR, Amy PS, Papelis L, Chen Y-T, Unz R. 2003. Final report: the fate and transport of perchlorate in a contaminated site in the Las Vegas Valley. Washington (DC): US Environmental Protection Agency, Office of Research and Development.

[EDSTAC] Endocrine Disruptor Screening and Testing Advisory Committee. 1998. Final report. Washington (DC): US Environmental Protection Agency, Office of Research and Development.

Erickson GE. 1983. The Chilean nitrate deposits. Am Sci 71:366–374.

Jackson WA, Rainwater K, Anderson TA, Lehman T, Tock R, Rajagopalan S, Ridley M. 2004. Distribution and potential sources of perchlorate in the high plains region of Texas. Austin (TX): Texas Commission on Environmental Quality.

Margolis S. 21 Jan 1986. Memorandum: health assessment, San Gabriel Valley. Washington (DC): Agency for Toxic Substances and Disease Registry, US Department of Health and Human Services.

Mayer KP. 2003. Memorandum: transmittal of perchlorate maps and table updated April 2003. San Francisco (CA): US Environmental Protection Agency Region 9.

National Research Council of the National Academies. 2005. Health Implications of Perchlorate Ingestion. Hetzel BS. 1989. The story of iodine deficiency: an international challenge in nutrition. Oxford (UK): Oxford Univ Pr.

Orris GJ, Harvey GJ, Tsui DT, Eldrige JE. 2003. Preliminary analyses of perchlorate in selected natural materials and their derivative products. USGS Open File Report 03-314.

Parsons, Inc. 2001. Scientific and technical report for perchlorate biotransport investigation: a study of perchlorate occurrence in selected ecosystems. Interim final report. Brooks Air Force Base (TX): Air Force Institute for Environment, Safety and Occupational Health Risk Analysis.

Schilt AA. 2003. Perchloric acid and perchlorates. 2nd ed. Columbus (OH): GF Smith Chemicals.

Schumacher JC, editor. 1960. Perchlorates: their properties, manufacture and uses. New York (NY): Reinhold.

Smith PN, Theodorakis CW, Anderson TA, Kendall RJ. 2002. Preliminary assessment of perchlorate in ecological receptors at the Longhorn Army Ammunition Plant (LHAAP), Karnack, Texas. Ecotoxicology 10:305–313.

[SNWA] Southern Nevada Water Authority. 2004. Report: low level perchlorate sampling results in the Colorado River system and Lake Mead. February 13, 2004. Las Vegas (NV): Peggy Roefer, Regional Water Quality Supervisor,

SNWA.

Snyder SA, Vanderford BJ. 2005. Trace analysis of bromate, chlorate, iodate, and perchlorate in natural and bottled waters. Environ Sci Technol 39: 4586–4953.

Takata K. 23 Dec 1985. Memorandum: request for CDC assistance regarding potential health effects of perchlorate contamination at the San Gabriel Valley Superfund sites. San Francisco (CA): US Environmental Protection Agency Region 9.

Urbansky ET. 1998. Perchlorate chemistry: implications for analysis and remediation. Bioremediat J 2:81–95.

Urbansky ET. 2002. Perchlorate as an environmental contaminant. Environ Sci Pollut Res Int 9:187–192.

Urbansky ET, Brown SK, Magnuson ML, Kielty CA. 2001. Perchlorate levels in samples of sodium nitrate fertilizer derived from Chilean caliche. Environ Pollut 112:299–302.

[USEPA] United States Environmental Protection Agency. Feb 1997. Special report on environmental endocrine disruption: an effects assessment and analysis. Washington (DC): USEPA. Office of Research and Development. EPA/630/R-96/012. 111 p. Available from: http://www.epa.gov/ORD/WebPubs/endocrine/. Accessed 23 Dec 2005.

[USEPA] United States Environmental Protection Agency. 11 May 1998. Memorandum: nationwide occurrence of perchlorate contamination. San Francisco (CA): Felicia Marcus, Region 9 Administrator.

two

Environmental Fate of Perchlorate

W Andrew Jackson, Todd A Anderson, Jaclyn E Cañas, Shane A Snyder, Kui Tan

Introduction

Perchlorate fate in the environment is largely controlled by the movement of water and biological transformation and translocation (Figure 2-1). Perchlorate has been generally viewed as being chemically or physically nonreactive under typical environmental conditions, although with respect to sorption processes this may be an unwarranted oversimplification. Adsorption did not appear to be significant in a variety of soils examined in a soil column study (Tipton et al. 2003), although an earlier study observed slight adsorption of perchlorate to variable charged soils at low pH conditions (Liang and Ling 1992). Certainly biological (microbial and plant) transformation largely controls the fate of perchlorate in most environments. In addition to transformation, plants can also effect the distribution of perchlorate by their large evapotranspiration potential and the subsequent effects on groundwater movement and potential re-release of perchlorate back to the environment. Most research conducted to date has involved ex situ remediation in which contaminated groundwater is pumped to the surface and treated by biological reactors (Kim and Logan 2000; Logan 2000, 2001; Logan and Lapoint 2002; Nerenberg et al. 2002), anion exchange (Gu et al. 2000, 2002), or other techniques (Urbansky 1998; Na et al. 2002). In addition, a number of sites have successfully implemented in-situ biological augmentation. Much less information is available on the fate (unaugmented biological transformation and plant uptake) of perchlorate in natural systems. This chapter reviews the current state of knowledge regarding the fate and transport of perchlorate in natural systems, specifi-

Perchlorate Ecotoxicology. Ronald J Kendall and Philip N Smith, editors.

ISBN-10: 1-880611-90-2 ISBN-13: 978-1-880611-90-6

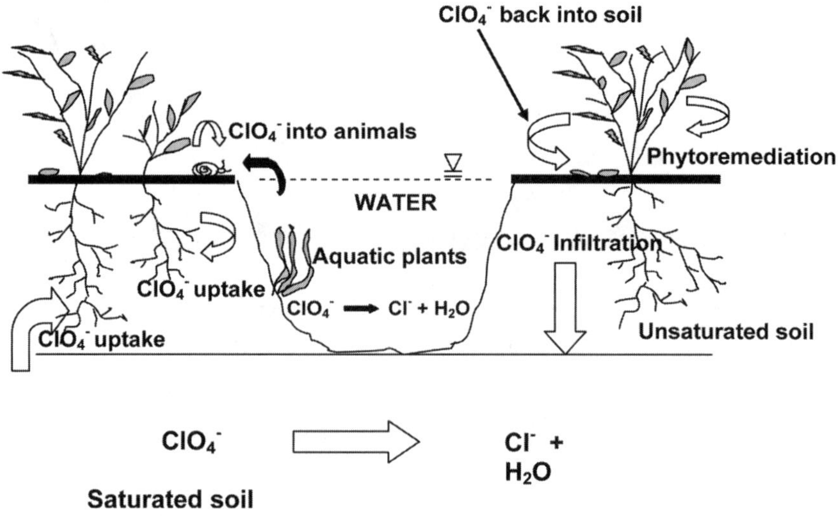

Figure 2-1 Perchlorate fate in the environment

cally focusing on the microbial and plant processes that control its fate and movement. In addition, a discussion of analytical issues dealing with detecting perchlorate in these diverse matrices is included.

Plant Uptake

Since 1999, a number of studies have been conducted that pertain to perchlorate accumulation in plants. Table 2-1 presents an overview of the known studies, including the plant species examined and whether they were field or laboratory studies. To date, there is a large pool of information on the uptake of perchlorate by non-crop aquatic and terrestrial plants, including parrot-feather, smartweed, pickleweed, sweet gum, water lily, and black willow (Susarla et al. 1999, 2000); salt cedar in Las Vegas Wash (Urbansky et al. 2000); tobacco plants (Ellington et al. 2001); bullrush, crabgrass, cupgrass, and goldenrod (Smith et al. 2002); smartweed, watercress, ash, chinaberry, Chinese elm, common paper mulberry, sugar hackberry, and willow (Tan et al. 2004a).

Perchlorate has also been detected in tobacco plants (*Nicotiana tabacum*) (Ellington et al. 2001) and food crops (*Cucumis sativus* L. and *Glycine max*) (Yu et al. 2004). Hutchinson et al. (2000) investigated the accumulation of perchlorate in lettuce irrigated with perchlorate-contaminated water, and in a recent study it was observed that about 18% of the supermarket lettuce from a total of 22 lettuce samples contained detectable amounts of perchlorate (EWG 2003).

Table 2-1 Summary of known studies on perchlorate accumulation in plants

Plant	Study summary	Reference
Willow (*Salix nigra*)	Hydroponic lab study	Nzengung et al. 1999
Parrot-feather (*Myriophyllum aquaticum*)	10-d exposure hydroponic	Susarla et al. 1999
Smartweed (*Polygonum* sp.)	10-d lab study; exposed to 20 ppm ClO_4^-	Susarla et al. 2000
Water-lily (*Nymphaea odorata*) Sweet gum (*Liquidambar styraciflua*) Willow (*Salix nigra*) Smartweed (*Polygonum punctatum*) Pickleweed (*Allenrolfea occidentalis*)	10-d exposure sand or hydroponic	Susarla et al. 2000
Salt cedar (*Tamarix ramosissima*)	Field study in Las Vegas Wash	Urbansky et al. 2000
Tobacco (*Nicotiana tabacum*)	Field study	Ellington et al. 2001
Bulrush (*Scirpus* sp.) Crabgrass (*Digitaria* sp.) Cupgrass (*Erichloa* sp.) Goldenrod (*Solidago* sp.)	Field study at Longhorn Army Ammunition Plant (LHAAP), Karnack, TX	Smith et al. 2001
Poplar (*Populus deltoidex nigra*)	Lab study using radiolabelled $^{36}ClO_4^-$	Aken and Schnoor 2002
Bulrush (*Scirpus* sp.)	Lab study	Tan et al. 2004b
Cucumber (*Cucumis sativus* L.) Soybean (*Glycine max*) Lettuce (*Lactuca sativa* L.)	Lab study	Yu et al. 2004
Smartweed (*Polygonum* sp.) Watercress (*Nasturtium* sp.) Willow (*Salix nigra*) Ash (*Fraxinus greggii*) Chinaberry (*Malia azedarach*) Elm (*Ulmus parvifolia*) Mulberry (*Broussonetia papyrifera*) Hackberry (*Celtis laevigata*)	Field study at Naval Weapons Industrial Reserve Plant (NWIRP), McGregor, TX	Tan et al. 2004a

These studies vary widely in their methodologies and purposes, but in general all of the studies showed the presence of perchlorate, primarily in above-ground plant tissue after some length of exposure. By far the most data available concern the accumulation in the leafy compartment of vegetation, but some information is available on perchlorate partitioning within plants. The mechanisms and rates of perchlorate uptake are the least understood aspects concerning plant uptake, although a (colloquial) data set exists on pseudo–steady-state bioconcentration factors (BCFs).

Uptake mechanisms

The uptake and ionic transport mechanisms of perchlorate in plants are not fully understood. Numerous physical and chemical properties, such as ionic concentration, electrochemical gradient, size (ionic or hydrated radius), and charge density, will affect the absorption and accumulation of anion solutes in plants. Often water and ionic transport through plants is through osmosis, but plants still absorb minerals against a higher concentration gradient by active transport mechanisms. The uptake of ions by plants involves a series of events. In general, the roots intercept the ions, which must then be taken up by the root and transported radially by 1 or more of a number of pathways such as uniport, cotransport, symport, or antiport. Finally, the ions destined for the shoots enter the leaf tissues.

The transport of solute anions (nutrients or contaminants) in plants is governed by membrane transport because the solute molecules have to pass the plasma membrane of root cells. Typically, membrane transport can be divided into passive (simple diffusion and facilitated diffusion) and active transport (Hopkins 1999). Nonpolar solute molecules (O_2, CO_2, and NH_3) appear to pass the lipid bilayer via simple diffusion caused by a concentration difference. Facilitated diffusion involves channel or carrier proteins but is still controlled by the concentration or electrochemical gradient. In active transport, ionic pumps (carrier proteins) are involved to transport ions against a concentration or electrochemical gradient. Active transport is reported for NO_3^-, $H_2PO_4^-$, and SO_4^{2-} in pea (*Pisum sativum*) and oat (*Avena sativa*) (Hopkins 1999). In fruits and seeds of plants, the phloem carries both inorganic and organic molecules (Salisbury and Ross 1992), and active transport is involved.

Recent data suggest that perchlorate is probably transported with water in xylem because perchlorate uptake appears to be in proportion with transpiration in smartweed grown in hydroponic systems. From unpublished data, NO_3^-, PO_4^{3-}, and SO_4^{2-} have no obvious effect on perchlorate uptake, implying that they are not sharing a common carrier protein.

These data suggest that passive transport may be involved in perchlorate transport in vascular plants. Passive transport is also supported by the low BCFs in non-leafy vegetation, which is dependent on perchlorate transport in phloem.

Equilibrium partitioning

The accumulation of perchlorate in plants could be affected by numerous potentially interrelated factors, such as concentration of perchlorate in bulk water (aquatic) or pore water (terrestrial), the length of exposure, type of plant, size of plant, structure of plant (ratio of leafy vegetation to non-leafy), solution composition, and potentially many more factors. Of these factors, substantial information is available on the effect of exposure concentration, length of exposure, and partitioning within plants.

Perchlorate exposure can be due to either contaminated soil or water; however, for perchlorate to be available to the plant, it must be in solution and in contact with a root system. Therefore, exposure concentrations are best expressed as solution concentrations because soil concentrations would require knowledge of the percent saturation and porosity in order to determine an exposure solution concentration. Exposure concentration could be expected to affect both the rate of uptake and the long-term plant concentration of perchlorate. Of these two, little information is available on the rate of uptake, but a relatively robust data set is available on long-term accumulation of perchlorate in leafy tissue.

Bioconcentration factor, the ratio between perchlorate concentration in plants ($mg \cdot kg^{-1}$ fresh weight or dry weight) and perchlorate concentration in media ($mg \cdot L^{-1}$), is commonly used to characterize the equilibrium between biota and water-phase contaminant concentration. It should be noted that concentrations can be expressed as $mg \cdot kg^{-1}$ in plants and $mg \cdot L^{-1}$ in water, resulting in a dimensionless concentration factor. A few lab and field studies have explicitly determined BCFs for perchlorate, and a number of other studies provide sufficient data for calculation of BCFs. In comparison to lab studies, which have reported BCFs for controlled exposures, BCFs (fresh weight) ranged from 170 to 620, while field studies have reported BCFs (fresh weight) from 215 to 380 for irrigated agricultural crops (Jackson et al. 2005). However, BCFs for non-crop species such as aquatic macrophytes (smartweed and water cress) range from 167 to 267 but on a dry weight basis (Tan et al. 2004a). These BCFs are generally lower than for irrigated crops, which are also expressed on a dry weight basis. The lower BCFs for the field studies are not surprising, given the likely variability in water sources for field samples (e.g., rain events), the lack of information on lateral and vertical perchlorate distribution, or in some cases a measure of the relative temporal variability in perchlorate concentrations. In addition, it is also possible that the lower field BCF is due to collection of younger plant material, which may not have reached true equilibrium. In laboratory studies, BCF values were calculated for plants harvested at maturity.

Length of exposure

The exposure duration is certainly a strong determinant of leaf concentration for short exposures (<3 weeks); however, the influence of long-term exposure is less clear. In a 1-y seasonal study of perchlorate uptake in deciduous trees bordering a stream contaminated with perchlorate, perchlorate concentrations in tree leaves generally increased throughout the growing season (Figure 2-2) (Tan et al. 2004a). The same tendency of perchlorate accumulation is seen in a willow tree sampled from the Naval Weapons Industrial Reserve Plant (NWIRP) site in central Texas, for which the increase appears to be consistent from year to year (Figure 2-3). However, in lab studies with common agricultural crops exposed to a 1-time dose of perchlorate (e.g., soybean, cucumber), plant concentrations peaked fairly rapidly (3 to 4 weeks) for cucumber, lettuce, and soybean and remained constant throughout the remaining growth cycle (generally <8 weeks) (Yu et al. 2004). Common complicating factors include the difference in total uptake versus plant concentration. Plants typically grow in spurts with the production of new leaves, so periods of rapid growth could result in large changes in tissue concentrations without major changes in total uptake. One possible explanation for the differences in field and lab studies in relation to the data from Tan et al. (2004a) may be due to the vastly different size of the plants studied. In the field study, the biomass density would have been much higher than in lab studies.

Rate of uptake

Up to now, the kinetic data on the rate of perchlorate uptake in plants have been scarce. The kinetic data reported so far were obtained by linear (zero-order) or nonlinear regression (1st-order decay) of the perchlorate concentration loss in growth medium (water, sand, or soil) versus time. The kinetic uptake experiments reported to date were relatively short (10 to 20 d), so the effects of long-term exposure (months or years) on the kinetic rates have not been elucidated. Susarla et al. (1999) reported perchlorate uptake rates that range from 0.004 to 0.090 h^{-1} for parrotfeather *(Myriophyllum aquaticum)*, assuming that plant uptake accounted for all of the perchlorate loss and ignoring microbial reduction. The 1st-order uptake rates for 6 vascular plant species (sweet gum, black willow,

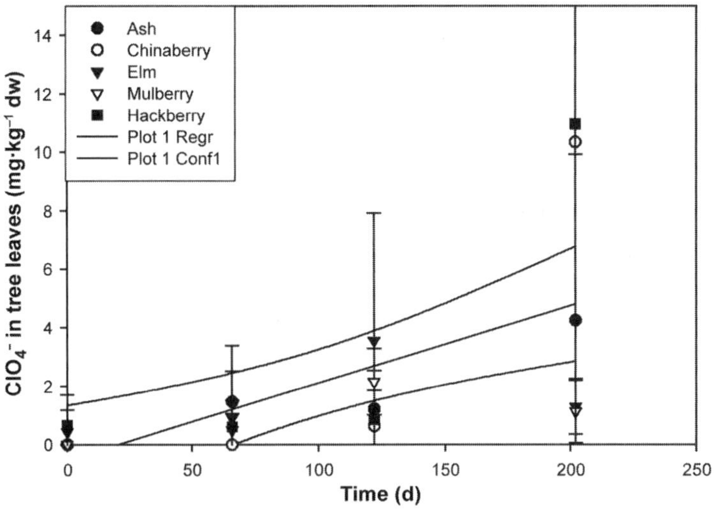

Figure 2-2 Perchlorate concentration in tree leaves bordering a stream during the growing cycle (data from Tan et al. 2004a)

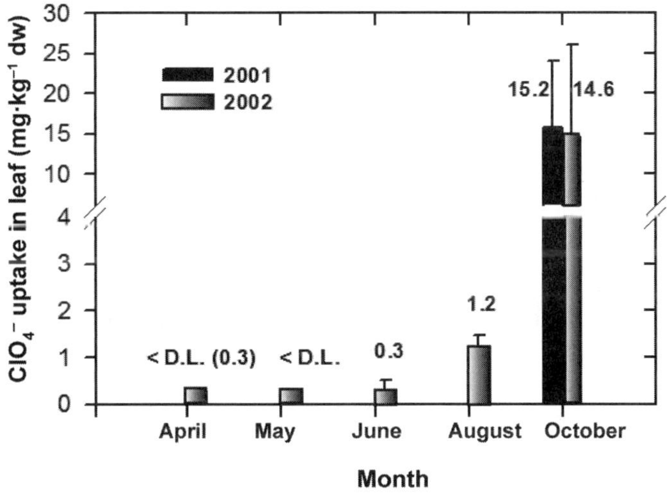

Figure 2-3 Perchlorate leaf seasonal accumulation in a willow tree grown at NWIRP, McGregor, Texas (data from Tan et al. 2004a)

pickleweed, smartweed, water lily, and duckmeet) ranged from 0.001 to 0.007 h^{-1} in a 10-d exposure laboratory experiment (Susarla et al. 2000). Nzengung et al. (1999) reported zero-order rate constants ranging from 0.003 to 2.35 $mgL^{-1}h^{-1}$ for willow trees growing in a variety of bioreactor conditions.

Partitioning within plant

Most previous studies (Susarla et al. 1999, 2000; Urbansky et al. 2000; Ellington et al. 2001; Smith et al. 2001; Tan et al. 2004a; Tan et al. 2004b) suggested that perchlorate was taken up and accumulated mainly in the aboveground portion of the plants, especially in the foliage. Perchlorate concentration in leaf tissues was 1 or 2 orders of magnitude higher than that in the water, implying a concentration and accumulation effect in the foliage. Recent studies indicate that perchlorate is accumulated in other edible parts of food crops such as fruits, seed, and grain, though with a relatively lower BCF compared to leaf tissue (Table 2-2). In a preliminary assessment at the Longhorn Army Ammunition Plant (LHAAP) near Karnack, Texas, extremely high concentrations of perchlorate were found in the blades and leaves of crabgrass, bulrush, goldenrod, and cupgrass. Perchlorate concentration was observed to be greatest in goldenrod leaves when compared to seeds stems and roots (Smith et al. 2001) (Table 2-3). Substantial exposure is possible for non-leafy vegetation. The accumulation of perchlorate in edible portions of forage and food crops is recently recognized as another possible exposure route to human and animal receptors in addition to drinking water.

Transformation potential

Perchlorate transformation in plant tissues has been examined in only a few studies (Nzengung et al. 1999; Aken and Schnoor 2002). Recently developed extraction, cleanup, and analytical methods make it possible to better understand the mechanisms involved in plant tissues. Radio-labeling techniques are commonly used to study the degradation pathway of contaminants, especially involving a low-level exposure. Phytodegradation of percholorate in plant tissues appears to occur fairly slowly, following a pathway of stepwise reduction ($ClO_4^- \rightarrow ClO_3^- \rightarrow ClO_2^- \rightarrow Cl^-$) (Nzengung et al. 1999; Aken and Schnoor 2002). Slow transformation in plant tissues may raise concerns over the final fate of perchlorate in deciduous tree leaves. By using willow trees grown in sand and hydroponic bioreactors, Nzengung et al. (1999) identified two important mechanisms involved in phytoremediation: 1) uptake and phytodegradation in plant tissues and 2) rhizodegradation (enhanced microbial degradation stimulated by

Table 2-2 Typical bioconcentration factor (BCF) in different tissues of food crops

Study	Plant type	BCF
Greenhouse[a]	Soybean leaves	620
	Soybean pod	150
	Soybean seeds	12
	Tomato leaves	220
	Tomato fruit	3.4
	Alfalfa	170
Field study[a]	Wheat stem	215
	Wheat head	233
	Alfalfa	380
	Cucumber (fruit)	5.7
	Cantaloupe (fruit)	20
	Tomato (fruit)	1.6

[a]Jackson et al. 2005

Table 2-3 Distribution of perchlorate in plants collected from different field sites

	Compartment ($\mu g \cdot kg^{-1}$ dw)		
Plant	Stem	Leaves	Fruit or seed
Crabgrass[a]		5560000	1880000
Goldenrod[a]	14000	1030000	184000
Blackberry[a]			593000
Chinaberry[b]		5000	ND
Mullberry[b]		1300	500
Wheat[c]	4100	1500	
Wheat[c]	1800	1500	
Wheat[c]	1600	9500	
Wheat[c]	17000	4000	

[a] Smith et al. 2001
[b] Tan et al. 2004b
[c] Jackson 2005

plant-released exudates and enzymes). Nzengung et al. (1999) also found that high NO_3^- concentrations interfered with rhizodegradation of ClO_4^-.

By using $^{36}ClO_4^-$, Aken and Schnoor (2002) provided further evidence of relatively slow stepwise ClO_4^- transformation in plant tissues.

Microbial Transformation of Perchlorate

Perchlorate-respiring bacteria are facultative anaerobes, and both heterotrophic and autotrophic forms have been isolated (Xu et al. 2003). Perchlorate, used as an electron acceptor, can be degraded by perchlorate reducers to innocuous Cl^- following the pathway of $ClO_4^- \rightarrow ClO_3^- \rightarrow ClO_2^- \rightarrow Cl^- + O_2$ (Rikken et al. 1996). Most research conducted to date has focused on ex situ reactors designed to remove perchlorate from contaminated groundwater or process streams (Kim and Logan 2000; Logan 2000, 2001; Logan and Lapoint 2002; Nerenberg et al. 2002). Fewer researchers have studied the microbial ecology of sites to understand the complex relationships between substrate, electron acceptors, vertical distribution, and seasonal effects (Wu et al. 2001; Tipton et al. 2003; Tan et al. 2004b). There are also numerous studies on the microbiology of perchlorate reducers, including the types of bacteria, commonly used substrates, and electron acceptors of identified organisms.

Species type and distribution

Perchlorate-reducing bacteria have been recognized to be ubiquitous, abundant, and diverse in a wide range of natural environments, including pristine and hydrocarbon-contaminated soils, sediments, sludges, and lagoons (Coates et al. 1998, 1999; Logan 1998, 2001; Wu et al. 2001). The population of perchlorate-reducing bacteria in the environment ranges from 2.3×10^3 to 2.4×10^6 cells per gram of sample (Coates et al. 1999). To date, more than 20 perchlorate-reducing bacteria have been isolated and characterized (Coates et al. 1999; Xu et al. 2003). Perchlorate-reducing bacteria isolated so far are Gram-negative and facultative anaerobes (Xu et al. 2003). The predominant bacteria appear to belong to the *Dechloromonas* and *Dechlorosoma* groups (Rikken et al. 1996; Bruce et al. 1999; Coates et al. 1999; Herman and Frankenberger 1999; Miller and Logan 2000; Achenbach et al. 2001; Logan et al. 2001; Zhang et al. 2002), although other bacteria groups have also been reported, such as *Vibrio dechloraticans* (Korenkov et al. 1976), *Wolinella succinogenes* (Wallace et al. 1996), *Citrobacter* (Okeke at al. 2002), and *Dechlorospirlium anomalous* (Michaelidou et al. 2000). The ubiquity and diversity of perchlorate-reducing bacteria in natural environments imply that microbial degradation is one of the most important fate processes involved in numerous perchlorate-contaminated sites. However, because these bacteria are all facultative anaerobes, the extent to which they transform perchlorate will largely be determined by the relative availabilities of organic matter and oxygen, as well as other environmental factors (saturated conditions.)

Role of alternative electron acceptors

Coupled with an electron donor, perchlorate is used as an electron acceptor, of which the last step is non–energy producing and is thought to be a detoxification mechanism. In general, the rates of transformation increase through the reaction, so intermediate products rarely accumulate (Giblin et al. 2000). The possible competition of other alternative electron acceptors has been well studied. The presence of high concentrations of oxygen will completely inhibit perchlorate reduction, but sulfate (SO_4^{2-}) and iron (Fe^{3+}) do not affect perchlorate reduction (Attaway and Smith 1993; Rikken et al. 1996; Herman and Frankenberger 1998, 1999; Coates et al. 1999). There are conflicting data on the effect of nitrate on perchlorate degradation (Attaway and Smith 1993; Rikken et al. 1996; Wallace et al. 1996). However, more evidence supports the hypothesis that the perchlorate reduction pathway and the denitrification pathway might not be correlated and separate terminal reductases might be involved (Coates et al. 1999; Giblin and Frankenberger 2001). Recent studies have shown that simultaneous perchlorate and nitrate removal can be achieved in the presence of sufficient substrate-electron donors (Xu et al. 2003; Tan et al. 2003; Tan et al. 2004b). Perchlorate degradation behavior and kinetics in natural sediments and soils are closely linked with the variation of organic substrate availability, bacterial population, and initial nitrate concentration (Tan et al. 2003).

Role of substrates

Both heterotrophic bacteria using organic carbon and autotrophic bacteria using hydrogen and bicarbonate have been isolated. Heterotrophic perchlorate reducers can grow on a variety of organic substrates, such as acetate, ethanol, butyrate, citrate, lactate, propionate, pyruvate, succinate, glucose, lactose, ethanol, methanol, peptone, yeast extract, casamion acids, and others (Logan 1998; Herman and Frankenberger 1999). Autotrophic bacteria utilizing hydrogen as an electron donor and bicarbonate as a carbon source to degrade perchlorate have also been reported (Giblin et al. 2000; Miller and Logan 2000; Nerenberg et al. 2002). In-situ bioremediation using a variety of electron donors, such as compost and cottonseed meal (Perlmutter et al. 2000); acetate (Cox et al. 2001); molasses and dried *Typha latifolia* stalk (Ives et al. 2002); poultry, cow manure, cotton waste, ethanol, molasses, and methanol (Kastner et al. 2001); and vegetable oil (Hunter 2002), have been tested or examined.

The concentration of substrates should meet the metabolism and energy requirements of perchlorate reducers. A molar ratio of acetate to ClO_4^- of 2.9 to 6.6 was used in packed-bed reactors to successfully remove perchlorate (Kim and Logan 2000), corresponding to a maximum $C:ClO_4^-$ weight ratio of 1.6:1. However, in the more complicated natural systems involving the presence of other competitive electron acceptors (i.e., nitrate, O_2), reduced bioavailability, as well as the use of a variety of substrates not as readily bioavailable as acetate, the concentration of substrates required for efficient in-situ bioremediation may be higher.

Rates of transformation

Microbial degradation in natural systems is commonly modeled as 1st-order kinetics. Intrinsic rates of perchlorate transformation in soil and sediment in the environment have been reported in only a few studies (Wu et al. 2001; Tan et al. 2003). Degradation rates ranged from 0.13 to 0.46 d^{-1}, and lag times of ClO_4^- degradation ranged from 0 to 60 d in multiple sediment and soil samples collected from the NWIRP, Texas site (Tan et al. 2004b). The presence of nitrate increased the lag time of perchlorate degradation, but higher organic substrate concentrations can reduce the lag time and enhance the degradation rates. Perchlorate degradation rates were also found in the range of 0.21 to 7.80 d^{-1} in an upflow wetland column study (Tan et al. 2004b).

Analytical Methods for Perchlorate Determination

The determination of perchlorate has been a challenge, especially in complex samples. A variety of techniques have been developed for the determination of perchlorate in water and non-water samples. These techniques each have notable strengths and weaknesses with detection limits ranging from 0.15 $\mu g \cdot L^{-1}$ to 640 $mg \cdot L^{-1}$ in water (Table 2-4).

Early techniques: colormetric and spectrophotometric

One of the first techniques for the quantitative determination of perchlorate was a colormetric method based on the formation of an insoluble complex of methylene blue with perchlorate (Nabar and Ramachandran 1959). The loss of methylene blue was determined spectrophotometrically and used to calculate the amount of perchlorate present. Other anions such as dichromate, sulfate, permanganate, molybdate, and periodate were

Table 2-4 Detection limits (ppb) for analytical techniques for perchlorate determination in various matrices

Method	Matrix	Detection limit	Reference
Colormetric	Water	NR	Nabar and Ramchandran 1959
Atomic absorption spectroscopy	Water	700 $\mu g \cdot L^{-1}$	Collinson and Boltz 1968
Fourier transform infrared (FTIR)	Water	640 000 $\mu g \cdot L^{-1}$	Kargosha et al. 1995
Spectroscopy	Fertilizers	200–700 $ng \cdot g^{-1}$	Collette et al. 2003
Capillary electrophoresis (CE)	Water	10 $\mu g \cdot L^{-1}$	Avdalovic et al. 1993
	Tobacco plants or products	300 $\mu g \cdot L^{-1}$	Ellington 2000
Raman spectroscopy	Water	100 $\mu g \cdot L^{-1}$	Kowalchyk et al. 1995
	Fertilizers	20 000 $ng \cdot g^{-1}$	Collette et al. 2003
Ion-selective electrodes (ISEs)	Drinking water	50–100 $\mu g \cdot L^{-1}$	Jezkova et al. 1997; Neuhold et al. 1996
	Pyrotechnic explosives	100 000 $ng \cdot g^{-1}$	Perez–Olmos et al. 1999
Gravimetry	Water	NR	Urbansky 1998
Flow injection analysis (FIA)	Salt samples	3 $ng \cdot g^{-1}$	Ensafi and Rezaei 1998
Electrospray ionization–mass spectrometry (ESI–MS)	Water, groundwater	1–10 $\mu g \cdot L^{-1}$	Handy et al. 2000; Koester et al. 2000; Magnuson and Urbansky 2000; Urbansky 2000
IC ESI MS	Wastewater	0.3 $\mu g \cdot L^{-1}$	Roehl et al. 2002
	Fertilizers	30 000 $ng \cdot g^{-1}$	Collette et al. 2003
Ion chromatography (IC)	Water	0.15–1000 $\mu g \cdot L^{-1}$	Maurino and Minero 1997; Wirt et al. 1998; Jackson et al. 1999, 2000; Okamoto et al. 1999; Liu et al. 2002
	Plants	5 $ng \cdot g^{-1}$	Ellington et al. 2001
	Tissues	7–700 $ng \cdot g^{-1}$	Anderson and Wu 2002; Narayanan et al. 2003
	Biological fluids	3–6 $\mu g \cdot L^{-1}$	Narayanan et al. 2003
	High-salinity water	0.77 $\mu g \cdot L^{-1}$	Tian 2003

found to also form precipitates with methylene blue, which interfered with perchlorate estimation via this method.

The next advancement in perchlorate analysis was the use of an atomic absorption spectrometric method (Collinson and Boltz 1968). This method was based on the formation of a complex of perchlorate with copper(I) that could then be analyzed via atomic absorption because of the presence of the copper.

Recent techniques

Fourier transform infrared spectroscopy

Within the last decade, Fourier transform infrared (FTIR) spectroscopy has been used in the analysis of perchlorate in water and hydroponic fertilizers (Kargosha et al. 1995; Collette et al. 2003). FTIR spectroscopy with an attenuated total reflectance crystal (ATR-FTIR) was used for perchlorate determination in hydroponic fertilizers. The ATR crystal probe was coated with a thin film of a water-soluble organometallic ion-exchange compound selective for perchlorate. This method provided a high degree of reproducibility, but a new coat must be placed on the ATR crystal for each analysis.

Capillary electrophoresis

Capillary electrophoresis (CE) with UV absorbance detection was used for the analysis of perchlorate in water and tobacco plants and tobacco products (Avdalovic et al. 1993; Ellington et al. 2001). Perchlorate displaces a highly UV-absorbing reagent in the CE buffer, which produces a negative peak that is then electronically reversed by the CE instrument. With this technique, perchlorate can be eluted well before other interfering ions such as bromide, chloride, and iodide.

Raman spectroscopy

Raman spectroscopy is not very sensitive but is a useful analytical tool because it can be used for samples with very high levels of total dissolved solids with only limited impact on perchlorate detection limits. Raman spectroscopy was used as an on-line detector for CE to detect sub-ppm concentrations of perchlorate in water (Kowalchyk et al. 1995) and provided detection limits of about 2 orders of magnitude below those for normal Raman spectroscopy. Normal Raman spectroscopy has also been used to detect perchlorate in fertilizers (Collette et al. 2003). One benefit of this method for perchlorate determination in fertilizers is that other sample components (nitrate, phosphate, sulfate, and urea) can be measured simultaneously with perchlorate while still maintaining adequate selectivity for perchlorate.

Ion-selective electrodes

Significant advances in the development of perchlorate ion-selective carbon paste and liquid/plastic electrodes have been occurring for a decade (Neuhold et al. 1996; Jezkova et al. 1997; Perez–Olmos et al. 1999). However, selectivity limits their applicability to solutions where perchlorate is known to be the dominant anion. Advantages of ion-selective electrodes (ISEs) include rapid response, sufficient selectivity, and sensitivity. However, interferences from other large anions are enough to render ISEs almost useless for typical drinking or raw waters (Urbansky 2000).

Flow injection analysis

A rapid, sensitive, and selective flow injection analysis (FIA) with spectrophotometric detection was developed for the determination of perchlorate (Ensafi and Rezaei 1998). Perchlorate in salt samples formed an extractable ion pair with Brilliant Cresyl Blue. The presence of iodide, nitrate, and chlorate interfered with the perchlorate analysis. However, these ions can be removed by evaporation of samples with concentrated hydrochloric acid.

Ion chromatography

Ion chromatography (IC) is currently the most common technique for the detection of perchlorate in water. There has been much advancement in perchlorate analysis by IC within the last few years (Maurino and Minero 1997; Wirt et al. 1998; Jackson et al. 1999, 2000; Okamoto et al. 1999; Liu et al. 2002). The recommended operating conditions, according to EPA Method 314.0, include a Dionex AS16 column, 1000 µL injection, hydroxide eluent and suppressed conductivity detection (Hautman et al. 1999). The analysis of perchlorate has not been limited to water samples. However, with a few notable exceptions (Ellington and Evans 2000; Anderson and Wu 2002; Kirk et al. 2003; Cheng et al. 2004), there has been little advancement in the analysis of biological samples. Most biological samples have been analyzed using the recommended EPA Method 314.0. for water. Slight modifications in sample extraction and cleanup (Ellington and Evans 2000; Anderson and Wu 2002; Cheng et al. 2004) have been used for the analysis of these biological samples. Recently, techniques that combine IC with electrospray ionization–mass spectrometry (ESI–MS) have been developed (Roehl et al. 2002; Collette et al. 2003).

The addition of an MS detector to an IC system provides for greater detection specificity and sensitivity.

An online preconcentration IC method was also developed for the determination of trace perchlorate in high-salinity water samples (Tian et al. 2003). Perchlorate was concentrated on a Dionex IonPac TAC-LP1 column. This system has been optimized for high-salinity water samples, but due to the ability of the method to remove general matrix interferences, Tian et al. (2003) suggested that this method could also be used for perchlorate analysis in biological tissue extracts. Subsequent studies supported this assertion (Kirk et al. 2003).

Mass spectrometry

Although IC with conductivity detection is widely used for routine monitoring of perchlorate, chromatographic retention times are not considered unique identifiers. Analytical methods have been developed for both water and biological samples, using liquid chromatography with mass spectrometric detection (LC–MS) for perchlorate analysis (Salov et al. 1992; Pantsar–Kallio and Manninen 1998; Urbansky et al. 1999; Handy et al. 2000; Magnuson et al. 2000). Although the liquid interface is generally ion chromatography, reverse phase chromatography with ion-pairing reagents has also applied (Urbansky et al. 1999; Urbansky 2000; Snyder et al. 2005). Mass spectrometric detectors offer the important advantage of selectivity because both retention time and molecular weight data are obtained. For instance, *p*-chlorobenzene sulfonate (PCBS) was found to be an interference with EPA Method 314 because it elutes at the same time as perchlorate and is conductive (Johnson et al. 2003). Mass spectrometry improves selectivity as the probability of a co-eluting compound of the same molecular weight as perchlorate (93 Daltons) is minimal. In addition, mass spectrometric analysis of perchlorate allows the simultaneous confirmation using the stable ^{37}chlorine isotope of perchlorate at 101 Daltons. Tandem mass spectrometry (MS–MS) utilizes molecular reaction monitoring (MRM) and has been applied to the analysis of perchlorate with $ng \cdot L^{-1}$ detection limits (Koester et al. 2000; Rickman 2003; Snyder et al. 2005; Winkler et al. 2004). For perchlorate, the 2 MRM transitions can be monitored, namely the loss of oxygen from the ^{35}perchlorate ion (99 Daltons transition to 83 Daltons) and the loss of oxygen from the ^{37}perchlorate ion (101 Daltons transition to 85 Daltons) (Rickman 2003; Snyder et al. 2005; Winkler et al. 2004). A major disadvantage of LC–MS has been matrix effects, namely ion suppression and ion enhancement, which negatively impact accurate quantitation (Handy et al. 2000; Koester et al. 2000; Gustavsson et al. 2001; Vanderford et al. 2003). For perchlorate, the most prevalent source of interference is sulfate; however, sulfate can be readily removed from aqueous samples using commercially available barium cartridges (Handy et al. 2000; Snyder et al. 2005; Winkler et al. 2004). A commercially available IC–MS is now available and is specifically marketed for sub-$\mu g \cdot L^{-1}$ quantitation of perchlorate. As MS equipment becomes more prevalent and of lower cost, it will likely become the preferred method for oxyhalide analysis.

Research Recommendations

The most critical needs for additional research relate to the uptake of perchlorate in plants. While a substantial data set is available, a number of data gaps should be addressed in future research. These gaps include the mechanism of uptake into the plant root and the associated molecular control systems, the rate of uptake and influencing factors, and more distribution data, especially for fruits. Also of interest is the continuing analytical development of methods for detection of perchlorate at low levels in difficult matrices (e.g., organs, plants). Finally, given the importance of the element iodine in relation to perchlorate's thyroid effect, better methods may be needed for analysis of the environmentally relevant forms of iodine as well as a better understanding of perchlorate uptake in plants in relation to iodine (all forms) uptake.

References

Achenbach LA, Michaelidou U, Bruce RA, Fryman J, Coates JD. 2001. Dechloromonas agitate gen. nov., sp nov and dechlorosoma suillum gen. nov., sp nov., two novel environmentally dominant (per)chlorate-reducing bacteria and their phylogenetic position. Int J Syst Evol Microbiol 51:527–533.

Aken BV, Schnoor JL. 2002. Evidence of perchlorate (ClO4–) reduction in plant tissues (poplar tree) using radio-labeled 36ClO_4^-. Environ Sci Technol 36:2783–2788.

Anderson TA, Wu TH. 2002. Extraction, cleanup, and analysis of the perchlorate anion in tissue samples. Bull Environ Contam Toxicol 68:684–691.

Attaway H, Smith M. 1993. Reduction of perchlorate by an anaerobic enrichment culture. J Ind Microbiol 12:408–412.

Avdalovic N, Pohl CA, Rocklin RD, Stillian JR. 1993. Determination of cations and anions by capillary electrophoresis combined with suppressed conductivity detection. Anal Chem 65:1470–1475.

Bruce RA, Achenbach LA, Coates JD. 1999. Reduction of (per)chlorate by a novel organism isolated from paper mill waste. Environ Microbiol 1:319.

Cheng Q, Perlmutter L, Smith PN, McMurry ST, Jackson WA, Anderson TA. 2004. A study on perchlorate exposure and absorption in beef cattle. J Agric Food Chem 52:3456–61.

Coates JD, Bruce RA, Haddock JD. 1998. Anoxic bioremediation of hydrocarbons. Nature 396:730.

Coates JD, Michaelidou U, Bruce RA, O' Connor SM, Crespi JN, Achenbach LA. 1999. Ubiquity and diversity of dissimilatory (per)chlorate-reducing bacteria. Appl Environ Microbiol 65:5234–5241.

Collette TW, Williams TL, Urbansky ET, Magnuson ML, Hebert GN, Strauss SH. 2003. Analysis of hydroponic fertilizer matrixes for perchlorate: comparison of analytical techniques. Analyst 128:88–97.

Collinson WJ, Boltz DF. 1968. Indirect spectrophotometric and atomic absorption spectrometric methods for determination of perchlorate. Anal Chem 40:1896–1898.

Cox EE, McMaster ML, Neville SL, Bonsack LT. 2001. Successful field demonstration of in situ bioremediation of perchlorate in groundwater [abstract]. In situ and on-situ bioremediation in the sixth international symposium; 4–7 Jun 2001; San Diego, CA.

Ellington JJ, Evans JJ. 2000. Determination of perchlorate at parts-per-billion levels in plants by ion chromatography. J Chromatogr A 898:193–199.

Ellington JJ, Wolfe NL, Garrison AW, Evans JJ, Avants JK, Teng Q. 2001. Determination of perchlorate in tobacco plants and tobacco products. Environ Sci Technol 35:3213–3218.

Ensafi AA, B Rezaei. 1998. Automatic liquid–liquid extraction flow injection analysis determination of trace amounts of perchlorate with spectrophotometric detection. Anal Lett 31:167–177.

[EWG] Environmental Working Group. 2003. Suspect salads: toxic rocket fuel found in samples of winter lettuce. Available from: http://www.ewg.org/reports/suspectsalads/es.php. Accessed 20 Jan 2006.

Giblin T, Frankenberger WTJ. 2001. Perchlorate and nitrate reductase activity in the perchlorate-respiring bacterium perclace. Microbiol Res 156:311–315.

Giblin TL, Herman DC, Frankenberger WT. 2000. Removal of perchlorate from ground water by hydrogen-utilizing bacteria. J Environ Qual 29:1057–1062

Gu B, Brown GM, Alexandratos SD, Ober R, Dale JA, Plant S. 2000. Efficient treatment of perchlorate contaminated groundwater by bifunctional anion exchange resins. In Urbansky ET, editor. Perchlorate in the environment. New York (NY): Kluwer Academic/Plenum. p 165–176.

Gu B, Ku Y–K, Brown GM. 2002. Treatment of perchlorate-contaminated water using highly-selective, regenerable ion-exchange technology: a pilot-scale demonstration. Remediation 12:51–68.

Gustavsson SA, Samskog JA, Markides KE, Longstrom B. 2001. Studies of signal suppression in liquid chromatography–electrospray ionization mass spectrometry using volatile ion-pairing reagents. J Chromatogr A 937:41–47.

Handy R, Barnett DA, Purves RW, Horlick G, Guevremont R. 2000. Determination of nanomolar of perchlorate in water by ESI-FAIMS-MS. J Anal At Spectrom 15:907–911.

Hautman DP, Munch DJ, Eaton AD, Haghani AW. 1999. Method 314.0. Determination of perchlorate in drinking water using ion chromatography, Revision 1.0. Cincinnati (OH): US Environmental Protection Agency. EPA Doc. No. 815-B-99-003.

Herman DC, Frankenberger WT. 1998. Reviews and analysis: Microbial-mediated reduction of perchlorate in groundwater. J Environ Qual 27:750–754.

Herman DC, Frankenberger WT. 1999. Bioremediation and biodegradation: bacterial reduction of perchlorate and nitrate in water. J Environ Qual 28:1018–1024.

Hopkins WG. 1999. Introduction to plant physiology, 2nd ed. New York (NY): J Wiley.

Hunter W. 2002. Bioremediation of chlorate or perchlorate contaminated water using permeable barriers containing vegetable oil. Curr Microbiol 45:287–292.

Hutchinson SL, Susarla S, Wolfe NL, McCutcheon SC. 2000. Perchlorate accumulation from contaminated irrigation water and fertilizer in leafy vegetables. In: Second International Conference on Remediation of Chlorinated and Recalcitrant Compounds; 2000 May; Monterey, CA.

Ives MA, Gearheart RA, Snible WJ. 2002. Isolation and performance of a wetland-associated perchlorate-reducing bacterium. 2nd International Conference on Wetlands and Remediation; 2001 Sep 05–06. Wetlands and Remediation II:93–100.

Jackson PE, Gokhale S, Streib T, Rohrer JS, Pohl CA. 2000. Improved method for the determination of trace perchlorate in ground and drinking waters by ion chromatography. J Chromatogr A 888:151–158.

Jackson PE, Laikhtman M, Rohrer JS. 1999. Determination of trace level perchlorate in drinking water and ground water by ion chromatography. J Chromatogr A 850:131–135.

Jackson WA, Joseph P, Laxman P, Tan K, Smith PN, Yu L, Anderson TA. 2005. Perchlorate in forage and edible vegetation: Concentrations in plants vs. contaminated irrigation water. J Agric Food Chem 53 (2) 369–373.

Jezkova J, Musilova J, Vytras K. 1997. Potentiometry with perchlorate and fluoroborate ion-selective carbon paste electrodes. Electroanalysis 9:1433–1436.

Johnson JD, Grimshaw D, Richman K. 2003. Analysis for perchlorate by ion chromatography: significant recent findings. Henderson (NV): American Pacific Corp 10.

Kargosha K, SH Davarani, Moosavi R. 1995. Quantitative analysis of binary aqueous sodium chlorate and sodium perchlorate solution in the presence of sodium dichromate using Fourier transform infrared spectrometry. Analyst 120:1945–1948.

Kastner JR, Das KC, Nzengung VA. 2001. In-situ bioremediation of perchlorate-contaminated soils. In: Leeson A, Peyton BM, Means JL, Magar VS, editors. Bioremediation of inorganic compounds. San Diego (CA): Battelle Pr. p 289–295.

Kim K, Logan BE. 2000. Fixed-bed bioreactor treating perchlorate-contaminated waters. Environ Eng Sci 17:257–265.

Kirk AB, Smith EE, Tian K, Anderson TA, Dasgupta PK. 2003. Perchlorate in milk. Environ Sci Technol 37:4979–81.

Koester CJ, Beller HB, Halden RU. 2000. Analysis of perchlorate in groundwater by electrospray ionization mass spectrometry/mass spectrometry. Environ Sci Technol 34:1862–1864.

Korenkov VN, Romanenko VI, Kuznetsov SI, Voronnov JV. 1976. Process for purification of industrial waste waters from perchlorates and chlorates. U.S. patent 3,943,055.

Kowalchyk WK, Walker III PA, Morris MD. 1995. Rapid normal Raman spectroscopy of sub-ppm oxy-anion solutions: the role of electrophoretic preconcentration. Appl Spectrosc 49:1183–1188.

Liang JG, Ling KX. 1992. Adsorption of chloride, nitrate and perchlorate by variable charge soils. Pedosphere 2:317–326.

Liu Y, Mou S, Heberling S. 2002. Determination of trace level bromate and perchlorate in drinking water by ion chromatography with an evaporative preconcentration technique. J Chromatogr A 956:85–91.

Logan BE. 1998. A review of chlorate- and perchlorate-respiring microorganisms. Bioremediat J 2:69–79.

Logan BE. 2000. Evaluation of biological reactors to degrade perchlorate to levels suitable for drinking water. In: Urbansky ET, editor. Perchlorate in the environment. New York (NY): Kluwer Academic/Plenum. p 189–197.

Logan BE. 2001. Assessing the outlook for perchlorate remediation. Environ Sci Technol 12:483A–487A.

Logan BE, Lapoint D. 2002. Treatment of perchlorate- and nitrate-contaminated groundwater in an autotrophic, gas phase, packed-bed bioreactor. Water Res 36:3647–3653.

Logan BE, Zhang HS, Mulvaney P, Milner MG, Head IM, Unz RF. 2001. Kinetics of perchlorate and chlorate-respiring bacteria. Appl Environ Microbiol 67:2499–2506.

Magnuson ML, Urbansky ET. 2000. Microscale extraction of perchlorate in drinking water with low level detections by electrospray-mass spectrometry. Talanta 52:285–291

Magnuson ML, Urbansky ET, Kelty CA. 2000. Determination of perchlorate at trace levels in drinking water by ion-pair extraction with electrospray ionization mass spectrometry. Anal Chem 72:25–29.

Maurino V, Minero C. 1997. Cyanuric acid-based eluent for suppressed anion chromatography. Anal Chem 69:3333–3338.

Michaelidou U, Achenbach LA, Coates JD. 2000. Isolation and characterization of two novel (per)chlorate-reducing bacteria from swine waste lagoons. In: Urbansky ET, editor. Perchlorate in the environment. New York (NY): Kluwer Academic/Plenum. p 271–283.

Miller JP, Logan BE. 2000. Sustained perchlorate degradation in an autotrophic, gas-phase, packed-bed bioreactor. Environ Sci Technol 34:3018–3022.

Na CZ, Cannon FS, Hagerup B. 2002. Perchlorate removal via iron-preloaded GAC and borohydride regeneration. J Am Water Work Assn 94:90–102.

Nabar GM, Ramachandran CR. 1959. Quantitative determination of perchlorate ion in solution. Anal Chem 31:263–265.

Narayanan L, Buttler GW, Yu KO, Mattie DR, Fisher JW. 2003. Sensitive high-performance liquid chromatrography method for the determination of low levels of perchlorate in biological samples. J Chromatogr B 788:393–400.

Nerenberg R, Rittmann BE, Najm I. 2002. Perchlorate reduction in a hydrogen-based membrane-biofilm reactor. J Am Water Work Assn 94:103–114.

Neuhold CG, Kalcher K, Cai X, Raber G. 1996. Catalytic determination of perchlorate using a modified carbon paste electrode. Anal Lett 29:1685–1704.

Nzengung VA, Wang CH, Harvey G. 1999. Plant-mediated transformation of perchlorate into chloride. Environ Sci Technol 33:1478–1479.

Okamoto HS, Dharmendra RK, Steeber WR, Baumann FJ, Perera SK. 1999. Using ion chromatography to detect perchlorate. J Am Water Works Assoc 91:73–84.

Okeke BC, Giblin T, Frankenberger WT. 2002. Reduction of perchlorate and nitrate by salt tolerant bacteria. Environ Pollut 118:357–363.

Pantsar–Kallio M, Manninen PKG. 1998. Speciation of halogenides and oxyhalogens by ion chromatography-inductively coupled plasma mass spectrometry. Anal Chim Acta 360:161–166.

Perez–Olmos R, Rios A, Martin MP, Lapa RAS, Lima JLFC. 1999. Construction and evaluation of ion selective electrodes for perchlorate with a summing operational amplifier: application to pyrotechnics mixtures analysis. Analyst 124:97–100.

Perlmutter MW, Britto R, Cowan JD, Patel M, Jacobes A, Logan B, Craig M. 2000. Bioremediation of perchlorate-contaminated groundwater at naval weapons reserve plant McGregor, Texas. In: National Defense Industrial Association Proceedings; 26th Environmental Symposium and Exhibition, Long Beach, CA, USA, 27–30, 2000.

Rickman JE. 2003. LC/MS/MS measures perchlorate in 0.1–0.3 ppb range. Trends Anal Chem 22:viii–ix.

Rikken GB, Kroon AGM, Ginkel CGV. 1996. Transformation of (per)chlorate into chloride by a newly isolated bacterium: reduction and dismutation.

Appl Microbiol Biot 45:420–426.

Roehl R, Slingsby R, Avdalovic N, Jackson PE. 2002. Applications of ion chromatography with electrospray mass spectrometric detection to the determination of environmental contaminants in water. J Chromatog A 956:245–254.

Salisbury FB, Ross CW. 1992. Plant physiology. 4th ed. Belmont (CA): Wadsworth.

Salov VV, Yoshinaga J, Shibata Y, Morita M. 1992. Determination of inorganic halogen species by liquid chromatography with inductively coupled argon plasma mass spectrometry. Anal Chem 64:2425–2428.

Smith PN, Theodorakis CW, Anderson TA, Kendall RJ. 2001. Preliminary assessment of perchlorate in ecological receptors at the Longhorn Army Ammunition Plant (LHAAP), Karnack, Texas. Ecotoxicology 10:305–313.

Snyder SA, Vanderford BJ, Rexing DJ. 2005. Trace analysis of bromate, chlorate, iodate, and perchlorate in natural and bottled waters. Environ Sci Technol 31 (12) 4586–4593.

Susarla S, Bacchus ST, Harvey G, McCutcheon SC. 2000. Phytotransformations of perchlorate contaminated waters. Environ Sci Technol 21:1055–1065.

Susarla S, Bacchus ST, Wolfe NL, McCutcheon SC. 1999. Phytotransformation of perchlorate and identification of metabolic products in Myriophyllum aquaticum. Int J Phytorem 1:97–107.

Tan K, Anderson TA, Jackson WA. 2003. Degradation kinetics of perchlorate in sediments and soils. Water Air Soil Pollut 151:245–259.

Tan K, Anderson TA, Jones MW, Smith PN, Jackson WA. 2004a. Accumulation of perchlorate in aquatic and terrestrial plants at a field scale. J Environ Qual 33:1638–1646

Tan K, Jackson WA, Anderson TA, Pardue JH. 2004b. Fate of perchlorate-contaminated water in upflow wetlands. Water Res 38:4173–4185.

Tian K, Dasgupta PK, Anderson TA. 2003. Determination of trace perchlorate in high-salinity water samples by ion chromatography with on-line preconcentration and preelution. Anal Chem 75:701–706.

Tipton DK, Rolston DE, Scow KM. 2003. Bioremediation and biodegradation: Transport and biodegradation of perchlorate in soils. J Environ Qual 32:40–46.

Urbansky ET. 1998. Perchlorate chemistry: implications for analysis and remediation. Bioremed J 2:81–95.

Urbansky ET. 2000. Quantitation of perchlorate ion: practices and advances applied to the analysis of common matrices. Crit Rev Anal Chem 30:311–343.

Urbansky ET, Magnuson ML, Freeman D, Jelks C. 1999. Quantitation of perchlorate ion by electrospray ionization mass spectrometry (ESI–MS) using stable association complexes with organic cations and bases to enhance selectivity. J Anal Atom Spectr 12:1861–1866.

Urbansky ET, Magnuson ML, Kelty CA, Brown SK. 2000. Perchlorate uptake by salt cedar (Tamarix ramosissima) in the Las Vegas Wash riparian ecosystem. Sci Total Environ 256:227–232.

Wallace W, Ward T, Breen A, Attaway H. 1996. Identification of an anaerobic bacterium which reduces perchlorate and chlorate as Wolinella succinogenes. J Ind Microbiol 16:68–72.

Winkler P, Minteer M, Willey J. 2004. Analysis of perchlorate in water and soil by electrospray LC/MS/MS. Anal Chem 76:469–473.

Wirt K, Laikhtman M, Rohrer J, Jackson PE. 1998. Low-level perchlorate analysis in drinking water and groundwater by ion chromatography. Amer Environ Lab 10:1–5.

Wu J, Unz RF, Zhang, Logan BE. 2001. Persistence of perchlorate and the relative numbers of perchlorate- and chlorate-respiring microorganisms in natural waters, soils, and wastewater. Bioremed J 5:119–130.

Xu JL, Song YG, Min B, Steinberg L, Logan BE. 2003. Microbial degradation of perchlorate: principles and applications. Environ Eng Sci 20:405–422.

Yu L, Canas JE, Cobb GP, Jackson WA, Anderson TA. 2004. Uptake of perchlorate in terrestrial plants. Ecotoxicol Environ Saf 58:44–49.

Zhang HS, Bruns MA, Logan BE. 2002. Perchlorate reduction by a novel chemolithoautotrophic, hydrogen-oxidizing bacterium. Environ Microbiol 4:570–576.

three

Thyroid Function and Mechanism of Perchlorate Action in Vertebrates

James A Carr, FM Anne McNabb, Ernest E Smith

Introduction

Several decades of clinical and laboratory research have elucidated the mode of perchlorate action, which is to block iodide (I^-) uptake by the thyroid gland. Because I^- is required for thyroid hormone (TH) synthesis, and THs act on virtually every tissue in the body to influence growth, differentiation, and metabolism, perchlorate meets the requirement for an endocrine disruptor as defined by the United States Environmental Protection Agency (USEPA 1997). Thus, it is important to understand the basic function of the thyroid gland and its regulation in order to predict potential health risks associated with perchlorate exposure in humans and wildlife.

In this chapter, we focus on the mechanism of perchlorate action on the mammalian thyroid gland, not to emphasize the importance of using mammals versus nonmammals for perchlorate studies, but because of the large data base that is available regarding basic thyroid function in mammals. The mammalian thyroid gland has long served as a model for vertebrate thyroid function, and many aspects of thyroid function are similar throughout the vertebrate classes, so some comparisons of perchlorate action can be drawn between mammals and nonmammalian vertebrates.

Perchlorate Ecotoxicology. Ronald J Kendall and Philip N Smith, editors.

ISBN-10: 1-880611-90-2 ISBN-13: 978-1-880611-90-6

Thyroid Function and its Hypothalamic–Pituitary Control

In mammals, the thyroid gland is a bi-lobed gland on the ventral surface of the larynx with the 2 lobes being connected by a narrow isthmus of epithelial and connective tissues. The gland produces 2 THs, tetraiodothyronine or thyroxine (T_4) and triiodiothyronine (T_3), which are involved in regulating many aspects of development and metabolism. Thyroid gland function is controlled by the hypothalamic–pituitary–thyroid (HPT) axis (Figure 3-1). The pituitary glycoprotein hormone, thyroid-stimulating hormone (TSH), acts on receptors in the thyroid gland to increase I^- uptake, TH synthesis and secretion, and thyroid gland growth. TSH is secreted from 1 of 5 phenotypically distinct cell types (thyrotropes) located in the anterior pituitary gland. Thyrotropes are controlled by hormones from the hypothalamus, primarily by a stimulatory tripeptide called thyrotropin-releasing hormone (TRH). TRH is produced by hypophysiotropic neurons in the paraventricular nucleus of the hypothalamus and is released into the portal vasculature of the median eminence, where it travels to the pituitary and acts on receptors located on thyrotropes. In addition to TRH, other hypothalamic hormones that also modulate pitu-

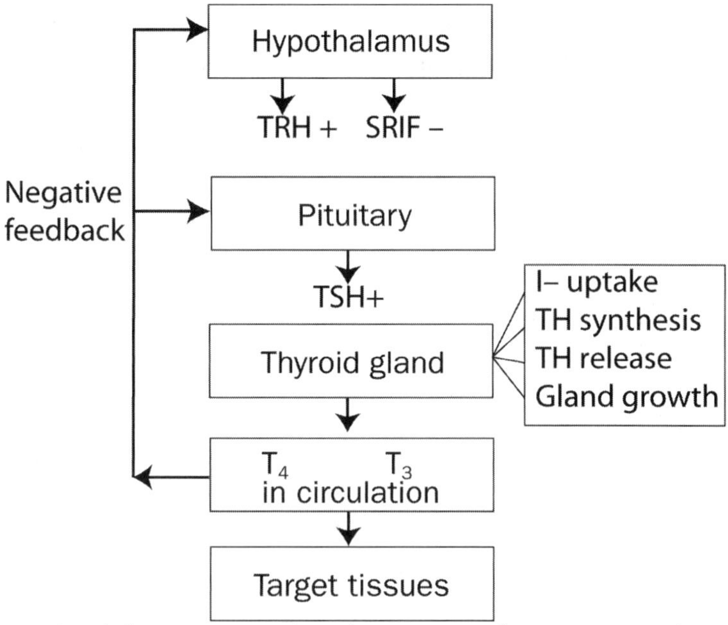

Figure 3-1 Hypothalamus–pituitary–thyroid axis in mammals, somatotropin-release inhibiting factor (SRIF), thyroid hormone (TH); thyrotropin-releasing hormone (TRH), thyroid-stimulating hormone (TSH, thyrotropin)

itary TSH secretion include somatotropin-releasing inhibiting hormone (SRIH), which inhibits TSH secretion, and in some species, corticotropin-releasing factor (CRF), which stimulates TSH secretion.

The production of TRH and TSH are negatively controlled by THs. Specifically, this feedback is due to T_3 binding to pituitary receptors. This T_3 comes both from the circulation and from T_4 monodeiodination within the pituitary (see "Metabolism of thyroid hormones," p 53). Feedback decreases the secretion of TRH (by a long feedback loop) and TSH (by a short feedback loop). Although perchlorate acts directly on the thyroid gland to inhibit I^- uptake and TH synthesis, an indirect consequence of reduced TH synthesis is activation of the HPT axis due to negative feedback. The indirect effects of perchlorate on negative feedback regulation of the HPT axis will be discussed in more detail later in the section titled "Clinical Disorders Related to Iodine Deficiency and Thyroid Hormone Synthesis." There are important differences in the organization of the HPT axis that should be taken into account when considering the effects of perchlorate on thyroid function in nonmammalian vertebrates. These comparative aspects of the HPT axis will be discussed in Chapters 5 through 8, which cover the effects of perchlorate on different vertebrate classes.

Production and secretion of thyroid hormones

In all vertebrates, THs are produced by cells called "thyrocytes" (or "follicle cells") that are organized into ball-shaped structures called "follicles." The wall of each follicle consists of a single-cell layer of thyrocytes that actively transport and accumulate I^- and synthesize the protein thyroglobulin (Tg), from within which THs are produced and stored (Figure 3-2). Although thyrocytes are best known for their role in TH synthesis, these cells share ultrastructural features in common with other absorptive epithelia, for example, the presence of tight intercellular junctions toward the apex of adjacent thyrocytes that play an important role in establishing the ionic gradients required for I^- transport. The hollow interior cavity of each thyroid follicle is filled with a protein-rich fluid called "colloid," within which Tg, containing THs formed from iodinated tyrosine residues, is stored. The thyroid gland is the only endocrine gland that stores its secretory products extracellularly and that contains sufficient stored hormone to supply needs for weeks or months.

THs are formed from 2 molecules of the amino acid tyrosine, with some carbons on the phenolic rings of each tyrosine forming covalent bonds with iodine atoms. Naming of THs is based on the location of the carbon

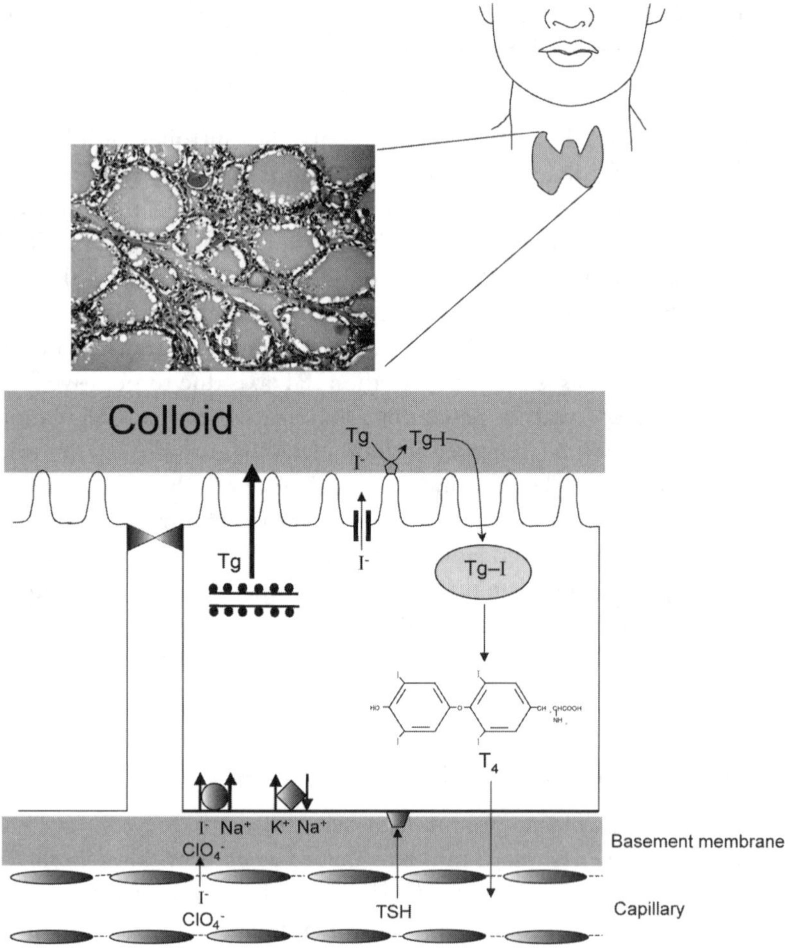

Figure 3-2 Basic steps in TH synthesis and secretion. The illustration shows a thyrocyte adjoined at the apical cell surface to an adjacent thyrocyte by tight or occluding junctions (zona occludens). Iodide (I–) and perchlorate (ClO_4^-) are transported into the thyrocyte by the sodium-dependent I^- symporter. The driving force for I^- transport is an inward sodium gradient established by the Na+/K+ ATPase. I^- is transported across the apical cell surface by pendrin or possibly other apical I^- transporters such as the human apical I^- transporter. In the follicle lumen, I^- is covalently bound to tyrosine residues on the protein thyroglobulin (Tg) by the enzyme thyroid peroxidase (TPO). TPO also conjugates adjacent tyrosine residues together to form thyronine molecules. Tg is taken up across the apical plasma membrane and proteolytically cleaved to release T_4, which enters the blood stream. Thyroid follicular epithelium produces a minimal basement membrane, ostensibly to enhance the efficiency of transport to and from the bloodstream. The pituitary glycoprotein hormone thyrotropin (TSH) acts on receptors located on the basal plasma membrane to stimulate I^- uptake and T_4 secretion.

molecules in the inner and outer carbon rings to which iodine is bound; by convention, the carbon ring that retains its alanine side chain is referred to as the "inner ring," while the other phenolic ring is called the "outer ring" (with carbon positions labeled with a prime designation). Thus, T_4 has iodine molecules attached at carbons 3, 5, 3′, and 5′, while T_3 lacks an iodine on the 5′ carbon. As shown in Figure 3-3, the phenolic ring structures of THs are linked by an ether bond, which renders THs relatively insoluble in water at neutral pH. Because iodine has a large atomic mass, the presence or absence of 1 iodine atom can dramatically affect the binding of TH to its receptor; in general, T_3 has 10-fold greater affinity for TH receptors than T_4, a key consideration in T_3 being considered the most metabolically active of the THs. Although T_3 is the active form of TH, almost all of the TH in thyroid glands of adult animals is T_4, which is converted in extrathyroidal tissues to T_3 via a process called "deiodination."

The process of TH synthesis consists of 3 basic steps:

- I^- uptake,
- coupling of I^- to tyrosine molecules within Tg to form iodothyronines (T_4 and T_3), and
- liberating the iodothyronine (i.e., TH) molecules from their protein scaffolding for release into the blood stream.

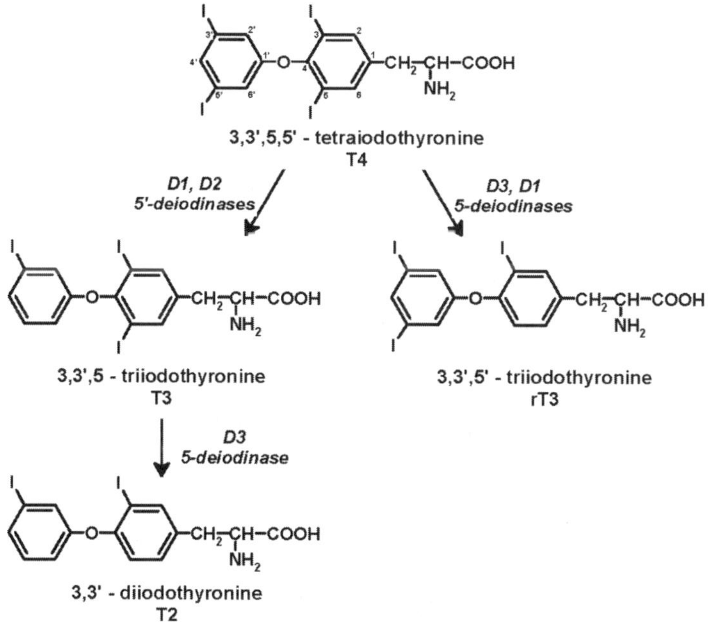

Figure 3-3 Structure and metabolism of thyroid hormones (with permission from Dr. Michael J Hooper, Texas Tech University)

Thyrocytes have evolved the capacity to selectively concentrate I^- from extracellular fluid containing relatively low I^- concentrations (10^{-8} M to 10^{-7} M). This I^- uptake occurs by a selective transport protein called the "sodium I^- symporter" (NIS), which transports both Na^+ and I^- across the basal plasma membrane and into the cytoplasm of the thyrocyte. This process is driven by a standing sodium concentration gradient across the basal plasma membrane, which is established by the sodium potassium ATPase (Na^+/K^+ ATPase) in the membrane. The Na^+/K^+ ATPase transports Na^+ out of the thyrocyte in exchange for K+ transport inward, thereby maintaining low intracellular Na^+ concentrations. As a result of this dependency on Na+ transport, I^- transport into thyrocytes can be disrupted by poisons, such as oubain, that interfere with the activity of the Na^+/K^+ ATPase. Although cells in other tissues such as gastric mucosa, salivary glands, and kidney tubular epithelium also express the NIS, thyrocytes express thyroid-specific transcription factors that promote NIS gene transcription (De La Vieja et al. 2001). In addition, the promoter region of the rat NIS gene has a TSH-responsive element that also up-regulates NIS expression (Ohmori et al. 1998). Recent work using TSH-receptor knockout mice indicates that NIS expression is virtually nonexistent in thyrocytes lacking TSH receptors (Marians et al. 2002).

As mentioned previously, thyrocytes are linked together by tight junctions, which separate the cells into functionally distinct apical and basal compartments. Although ATP-dependent transport of I^- from the blood stream clearly falls within the domain of NIS proteins in the basolateral compartment of thyrocytes, recent studies have identified two I^- transport proteins in the apical plasma membrane of thyrocytes. The discovery of apical I^- transporters was predicted because of the requirement for I^- to move across the apical plasma membrane into the lumen of the thyroid follicle in order for it to be incorporated into thyroid hormones. The first of these apical I^- transporters to be cloned, pendrin, is an 86 kDa protein encoded by the Pendred syndrome gene (PDS) (Everett et al. 1997; Royaux et al. 2000). Patients diagnosed with Pendred syndrome display mutations in the PDS gene, develop goiters, and exhibit problems with I^- organification (see next page). Pendrin is generally considered to be an anion exchanger, chaperoning I^- across the apical plasma membrane and into the follicle lumen in exchange for Cl^- (Yoshida et al. 2004). A second putative I^- transporter, the human apical iodide transporter (hAIT, Rodriguez et al. 2002), may also be involved in shuttling I^- into the follicle lumen, although this protein appears to be expressed in only a fraction of thyrocytes.

The tyrosines used to synthesize THs are within the amino acid "backbone" of Tg, which is a large (660 kDa), tyrosine-rich protein that is synthesized by thyrocytes. In mice, for example, each Tg monomer contains about 67 tyrosine molecules, and the protein is apparently synthesized independently of TSH control (Marians et al. 2002). In order for TH synthesis to occur, Tg must be transported into the follicle lumen, a process that occurs by an as of yet unidentified transport mechanism. In the follicular lumen, Tg acts as the scaffolding upon which tyrosine molecules are iodinated and coupled to form iodothyronines, and thus it is within Tg that THs are stored in the colloid, with Tg comprising up to 30% of the total colloidal protein (see review by Dunn and Dunn 2000). Subsequently, Tg is internalized into thyrocytes by the process of endocytosis and proteolytically degraded to release THs (see below). Thus, depletion of follicular colloid is generally associated with a decrease in stored THs. Measurements of colloid depletion have been used as indicators of TH depletion in mammals and nonmammalian vertebrates, including that caused by perchlorate (Siglin et al. 2000; Carr et al. 2003; Patiño et al. 2003; York et al. 2003).

Before it is covalently attached to tyrosine residues in Tg, I^- must first be oxidized. Oxidation of I^- is carried out by the enzyme thyroid peroxidase (TPO), which also incorporates the oxidized I^- into Tg and couples the iodinated tyrosines to form iodothyronines. The iodination of Tg occurs on the apical surface of thyrocytes, where hydrogen peroxide, which acts as an electron acceptor for the enzymatic reaction, is produced. Although there is sufficient cytoplasmic I^- and Tg to carry out organification steps intracellularly, it is a lack of cytoplasmic H_2O_2 (which is rapidly degraded by glutathione) that prevents iodination from occurring within thyrocytes (Nilsson 2001). TPO is the target for goitrogenic compounds such as propylthiouracil (PTU) and methimazole, which directly inhibit the organification of I^- and disrupt the coupling of iodinated tyrosine molecules in Tg (Meier and Burger 2000).

In order for TH to be released from the thyroid gland, Tg must be retrieved from the colloid by endocytosis at the apical membrane of thyrocytes and proteolytically cleaved in lysosomes. After degradation of Tg in lysosomes, THs are liberated and released into the bloodstream. Mono- and di-iodotyrosines released by the proteolysis of Tg are recovered by a specific transporter in lysosomal membranes, deiodinated by an iodotyrosine deiodinase in the cytosol, and recycled as inorganic iodine and tyrosine. Thyrocytes possess well-developed microvilli to increase absorptive surface area of the apical plasma membrane. Tg is retrieved from the colloid by fluid-phase pinocytosis as well as by 2 forms of receptor-mediated endocytosis, one of which involves low-affinity endocytic receptors and another that involves the low-density lipoprotein endocytic receptor megalin (see Marino et al. 2001 for review). Interestingly, some of the Tg that is internalized by megalin bypasses lysosomal degradation pathways, is transported across the thyrocytes (transcytosis) and is released into the blood stream (Lisi et al. 2003). In addition, some of the Tg present in colloid is not soluble and must undergo extracellular proteolysis in the colloid before it can undergo endocytosis and enter thyrocytes. Endocytosis of colloid (which leads to hormone release) in thyrocytes is stimulated by TSH.

Transport of thyroid hormones in the circulation

THs circulate in blood and some extracellular compartments attached to binding proteins with differing hormone affinities. Historically, these binding proteins were thought to be a means of solubilizing these relatively hydrophobic hormones, but more recently they have been considered to function as a store of hormones readily accessible to target tissues and to help regulate hormone distribution within different organs (review, Schreiber 2002). An extremely small fraction (<<1%) of the total TH in blood exists in free (i.e., unbound) form. There has been general support for the "free hormone hypothesis," which states that the free hormone fraction is the part available for entry into tissues and that the free hormone fraction thus plays a key role in hormone supply to, or effects on, target tissues (Robbins 1996). The affinity for THs, the on–off rate (i.e., the rate of TH binding or release), and the rate of blood flow in capillaries all play roles in TH availability to tissues (Schreiber 2002). However, recent work indicates that THs are transported into tissues by carrier-mediated or facilitated diffusion. This work is leading to a reevaluation of the free hormone hypothesis, which was at least partially based on the assumption that THs enter cells by diffusion (review, Hennemann et al. 2001).

Large mammals, including humans, have 3 types of TH-binding proteins, ranked from highest to lowest in their binding affinity for T_4:

- thyroxine-binding globulin (TBG),
- transthyretin (TTR; also called "thyroxine-binding prealbumin"),
- and albumin.

In humans, for example, TBG binds about 70% and TTR binds about 10% to 15% of circulating T_4 (Robbins 1996). Small mammals, birds, reptiles, amphibians, and fish lack TBG, so the principal TH-binding protein in these animals is TTR (Larsson et al. 1985; Schreiber 2002).

Binding proteins in the circulation are generally produced in the liver, whereas in those animals in which the choroid plexus produces TTR, this TTR is primarily supplied to the brain (Prapunpoj et al. 2002; Schreiber 2002). Mammalian TBG and TTR bind T_4 with higher affinity than T_3. In contrast, in the birds, crocodiles, and amphibians studied to date, TTR has higher affinity for T_3 than T_4 (Chang et al. 1999; Prapunpoj et al. 2002; Yamauchi et al. 1993, respectively). Studies of the evolution of TTR in the vertebrate classes indicate that TTR expression evolved first in choroid plexus and later in liver (Prapunpoj et al. 2002). Thus, differences in the tissue distribution of TTR and of its concentration in the circulation may play differential roles in thyroid dynamics in different classes of vertebrates.

The cells of the choroid plexus synthesize TTR, and TTR has been implicated in the transport of THs across the choroid plexus and into the cerebrospinal fluid (review, Schreiber 2002). However, studies in TTR knockout mice indicate that this protein is not required for TH transport into brain tissue (Palha et al. 2000). Recent studies have provided evidence that THs cross plasma membranes of other organs or tissues by energy-using transport mechanisms that play a role in regulating tissue hormone supply. Two categories of molecules that appear to transport THs are organic anion transporters and amino acid transporters. Current research in these areas is leading to a reevaluation of previous ideas about tissue TH supply (reviews, Hennemann et al. 2001; Friesema et al. 2005).

Metabolism of thyroid hormones

The metabolism of THs involves deiodination reactions, which can activate T_4 or deactivate either T_4 or T_3, and other degradative enzymatic pathways (nondeiodinative) that alter the hormones in ways that facilitate their excretion. The latter reactions involve conjugations (glucuronidation and sulfation), side-chain alterations, and cleavage of the ether bond between the 2 phenolic rings.

Deiodination of thyroid hormones

Almost all of the hormone produced by the thyroid gland is T_4, and most of the T_3 present in the circulation (at least 80% in humans; Engler and Burger 1984) is produced in extrathyroidal tissues (primarily liver and kidney) by the 5′deiodination of T_4. Thus, deiodination reactions can activate T_4 by converting it to T_3, which is the metabolically active form for which thyroid receptors have the highest affinity (see discussion of thyroid receptors in “Thyroid hormone receptors”). Deiodination reactions also can deactivate T_4 and T_3 by converting them to forms that do not have biological activity. The enzymes involved in deiodination are selenoproteins, and in recent years molecular investigations of them have greatly extended knowledge of their physiological and biochemical roles (reviews, Leonard and Koehrle 2000; Bianco et al. 2002).

Three deiodinases have been characterized in mammals. Their tissue distributions and the reactions they catalyze are as follows.

- Type I 5′deiodinase (D1) is found in many tissues, including liver, kidney, central nervous system, pituitary, thyroid gland, placenta, and intestine. D1 deiodinates T_4 into T_3 by outer ring deiodination (activation) or into reverse-T_3 (rT_3) by inner ring deiodination (deactivation) and is the only deiodinase that can act on either the outer or inner ring of T_4.
- Type II 5′deiodinase (D2) is found primarily in the pituitary, the central nervous system, and brown adipose tissue, and it deiodinates T_4 to T_3 (activation). Smaller amounts of D2 expression have been found in other tissues, and there are a number of species and class differences in D2 distribution in vertebrates.
- Type III 5 deiodinase (D3), which is found in the central nervous system, placenta, and embryonic liver, deiodinates the inner ring of T_4 to produce rT_3 and the inner ring of T_3 to produce diiodothyronine (both deactivations; Leonard and Koehrle 2000).

Overall, these enzyme systems play a variety of important roles in regulating the supply of T_3, the most metabolically active TH. The main source of T_3 supply to the circulation is from T_4 to T_3 conversion by D1 in the liver and kidney. The rate of D1 synthesis is influenced positively by TH, is decreased by selenium deficiency and fasting, and is lower in fetal than in postnatal and adult tissues. The D2 enzyme appears to play its most important role in the regulation of central nervous system T_3 within narrow limits, despite alterations in organismal thyroid status. This regulation is achieved by D2 responding inversely to organismal thyroid status (i.e., it is negatively regulated by circulating TH concentrations), which thereby protects or buffers the central nervous system from potentially adverse alterations in peripheral TH concentrations (Bianco et al. 2002). A second important function of D2 is in the feedback regulation of TSH secretion in the anterior pituitary, where it is responsible for producing about half of the T_3 that binds to pituitary TRs (thyroid hormone receptors). This feedback regulation involves T_3 from both the circulation and from intrapituitary deiodination (Silva et al. 1978). Third, D2 is important in generating intra-tissue T_3 that plays a key role in regulatory thermogenesis in brown adipose tissue (BAT) (see "Thyroid hormone effects on organismal metabolism and heat production"). Finally, the many differences in D2 distribution that have been identified argue for differential species-specific functions of this deiodinase.

In contrast to D1 and D2, the D3 is a deactivating enzyme that converts T_3 to an inactive diiodothyronine and that prevents the activation of T_4 by converting it to inactive rT_3 (both inner-ring deiodinations). Liver D3 activity is important in maintaining low circulating T_3 in mammalian and avian embryos and in some premetamorphic stages of amphibian development (reviews, Leonard and Koehrle 2000; Bianco et al. 2002).

Nondeiodinative TH degradation

In addition to the well-investigated deiodinations of THs, ether bond cleavage, side-chain degradation and conjugations (glucuronidation and sulfation) also may influence TH dynamics. All of these reactions are less well understood than deiodinations, and the physiological significance of ether bond cleavage (which yields mono- and di-iodotyrosines) is unclear. Metabolism of the alanine side chain of THs involves oxidative deamination. Of the compounds formed in this way, the acetic acid derivatives have some thyromimetic activity and appear to have received the most attention. However, these compounds have short half-lives and generally do not appear to play an important role in thyroid function and dynamics (Leonard and Koehrle 2000).

The 2 key conjugation reactions involving THs are glucuronidation (by uridine diphosphate glucuronosyltransferases [UDP-GT]), which occurs preferentially with T_4 because of its 2 iodines in the phenolic ring, and sulfation (by sulfotransferases), which occurs preferentially with T_3 because of its single iodine in the phenolic ring. It should be noted that both of these enzymes can act on both T_4 and T_3, but different isozymes have differential preferences for T_4 versus T_3. In addition to T_4 and T_3, some other iodothyronines are conjugated but those reactions involving hormonally inactive forms will not be discussed here. In general, these hepatic Phase II conjugations make thyroid hormones more hydrophilic and thereby facilitate their excretion into the gastrointestinal tract in bile. In the gastrointestinal tract, glucuronidated THs can be metabolized by intestinal bacteria, leading to the release of free iodothyronines that may be reabsorbed and thereby participate in enterohepatic cycling (DiStefano 1988).

Sulfation of THs leads to preferential deiodination, and in the case of T_4, deiodination and desulfation can result in activation, that is, T_3 production (Visser et al. 1998; Leonard and Koehrle 2000). Regarding the roles of conjugation reactions in the normal regulation of THs, sulfation has received the most attention because of the interactions between these pathways and deiodination reactions. In contrast, most studies of glucuronidation have focused on the effects of some pharmaceuticals and environmental chemicals (such as some polyhalogenated aromatic hydocarbons) that may induce conjugation enzymes, thereby enhancing TH excretion and disrupting thyroid homeostasis (Barter and Klaassen 1992a, 1992b, 1994; Brouwer et al. 1998; Leonard and Koehrle 2000).

Thyroid hormone function

Thyroid hormone entry into target tissues

Given the hydrophobic nature of THs, it was thought for a long time that these hormones freely diffuse through the plasma membrane of target cells. However, a large number of recent studies indicate that THs rely on carrier-mediated or facilitated diffusion to enter target cells (Friesma et al. 2005). A number of candidate TH-transport proteins, which may move THs across the plasma membranes of different cell types, have been identified. These include, for example, TTR, which may be involved in TH transport across the choroid plexus (see "Transport of thyroid hormones in the circulation," p 52) and the organic anion-transporting polypeptide (OATP) family (Hennemann et al. 2001; Pizzagalli et al. 2002). In particular, monocarboxylate transporter 8 (MCT8; Friesma et al. 2003) appears to be important in TH transport into the brain. Mutations in the MCT8 gene have been reported to be associated with X-linked psychomotor retardation in humans (Dumitrescu et al. 2004; Heuer et al. 2005). In general, it now appears that THs probably are transported across plasma membranes in most cell types, that there are a number of different transporters that may differ with tissue type, and that these transporters may play very different regulatory roles in different tissues or organs (Hennemann et al. 2001). This currently is a very active area of research, and many questions remain about how alterations in thyroid status are related to these transport dynamics from tissue to tissue within an organism and across different vertebrate classes. In addition, it is not clear how new information about plasma membrane transporters of THs will mesh with the "free hormone hypothesis," which suggests that tissue TH entry is based on the concentration of free (i.e., unbound) THs in the circulation.

Thyroid hormone receptors

Once inside the target cells, THs act on nuclear receptors that alter gene transcription and protein synthesis; in effect, thyroid hormone receptors

(TRs) act as transcription factors, and molecular studies have revealed that TR genes belong to a large family of nuclear receptors that act as transcription factors. Virtually all tissues respond to THs and express one or more TRs. Although T_3 is the endogenous ligand with greatest affinity for TRs, these receptors also will bind T_4 albeit with lower affinity, so T_4 may be responsible for some of the biological effects of THs. There are 2 distinct genes that encode TRs: the alpha gene and the beta gene. In addition, each TR gene produces mRNA splice variants, termed the TRα1, TRα2, TRβ1, and TRβ2, that have some differences in their tissue distribution and functions. TRs are similar to other nuclear hormone receptors or transcription factors in that each receptor protein has multiple functional domains:

- the A/B domain at the N terminus is responsible for transcription activation;
- the C domain is responsible for DNA binding;
- the D domain is believed to function as a molecular hinge, and
- the E domain at the C terminus is important for ligand (T_3) binding.

TRs interact with co-factors (or coactivators) to facilitate gene transcription. TRs have different functions in the absence of ligand (T_3). In the unliganded state, TRs associate with proteins called "co-repressors" that negatively influence transcriptional pathways controlling gene expression (Zhang and Lazar 2000).

Thyroid hormone effects on development

THs are required for the normal differentiation and growth of many tissue types, including skeletal and muscle tissue, the cardiovascular system, the central nervous system (CNS, see below), and the gastrointestinal tract. THs are important for the morphogenesis of cartilage into endochondral bone in the skeletal system (Bassett and Williams 2003), for normal development of the heart, and for the expression of specific isoforms of myosin and actin in muscle (Legrand 1986). The effects of THs on muscle differentiation are not well understood, and they involve interactions with growth hormone and growth factors (Scanes et al. 1986; Forhead et al. 2002) as well as with the retinoic acid receptor (Cassar-Malek et al. 1996). Many of the studies of TH effects on muscle have been on nonmammalian models (e.g., amphibians, Das et al. 2002; birds, Cassar-Malek et al. 1996; Bishop et al. 2000).

TH deficiency during postnatal development clearly has adverse affects on the heart (Book et al. 1997; Wickenden et al. 1997; Merkulova et al.

2000). Neonatal hypothyroidism retards protein synthesis in the heart, causing smaller hearts and lowered heart weights (Breall et al. 1984; Moussavi et al. 1985; Canavan et al. 1993). Breall et al. (1984) demonstrated that hypothyroidism during gestation and continuing through postnatal development lowers ventricular ejection, systemic blood flow, and oxygen consumption. In amphibians, THs have dramatic effects on development and are required for restructuring of the gastrointestinal tract, limb growth, tail resorption, and changes in many physiological systems that are associated with the transition from a tadpole to a frog (see Chapter 6).

One of the key features of TH effects on development is that many of these effects require only small amounts of TH for brief periods during critical stages in development. For example, if THs are not present during critical windows in brain development, many aspects of the CNS will not develop properly, leading to permanent deficiencies in function. Thus, there is the potential that contaminants (such as perchlorate) that disrupt TH synthesis need only be present during these short but critical periods of development in order to permanently alter developing organ systems.

THs are required for normal brain development, and deficiencies in TH during the perinatal period can result in severe and permanent deficiencies in many aspects of nervous control functions, including cognitive function in humans. The data supporting a role for TH in brain development have been extensively reviewed elsewhere (Porterfield and Hendrich 1993; Howdeshell 2002; Zoeller et al. 2002). By acting on TRs in the developing nervous system, THs affect virtually every aspect of neuron and glial growth, including development of axons, formation of synapses, production of myelin by oligodendrocytes and Schwann cells, and migrations of nerve cells. However, the exact mechanisms whereby inadequate TH supply alters many of these processes are not well understood and are under active investigation (see, e.g., the reviews by Bernal 2002 and Anderson et al. 2003). Experimentally induced neonatal hypothyroidism in laboratory animals results in reduced neuronal cell size and reduced myelination in cognitive, visual, and motor areas of the cortex and the cerebellum, leading to severe cognitive and motor deficits in developing animals. Lack of normal TH synthesis also has been linked to a genetic form of epilepsy in rats (Mills and Savage 1988). In rats, which are altricial in their development, many of the CNS deficits observed with TH deficiency during the perinatal period can be reversed by TH replacement, but only if given during critical windows in development. In humans, there remains a great deal of concern for infants exposed to even slight hypothyroidism in utero. Humans are more precocial than rats, so some critical stages of brain development occur prenatally. Children of women who were hypothyroid during pregnancy exhibit greater rates of mental retardation, a general lowering of IQ scores (Haddow et al. 1999), and impaired psychomotor development (Pop et al. 1999).

Thyroid hormone effects on organismal metabolism and heat production

The component of basal or resting metabolism that distinguishes homeothermic from poikilothermic vertebrates, obligatory metabolism, is considered to be under the control of THs. In addition, THs also play a role in the control of adaptive or regulatory heat production that occurs in response to cold (Danforth and Berger 1984). THs elevate in-vitro oxygen consumption in all tissues of mammals, except for brain, testes, and spleen. The effects of THs on cellular respiration or metabolism are complex and can be broken down into long- and short-term effects. Long term effects of THs on cellular respiration include alterations in genomic (versus mitochondrial) gene transcription and subsequent synthesis of mitochondrial proteins involved in the respiratory pathway (oxidative phosphorylation) as well as those that have possible effects on the lipid composition of mitochondrial membranes. Some short-term effects of THs are not well understood, especially those that appear to have direct actions on putative TRs in the mitochondria. Recent studies have identified a form of TRα in the inner mitochondrial membrane, and this suggests a mechanism for an extranuclear action of T3 in thermogenesis (Wrutniak-Cabello et al. 2001).

In addition to their role in obligatory heat production, THs are important in regulatory heat production and cold adaptation of mammals in response to low temperatures. The induction of nonshivering thermogenesis in BAT is initiated by sympathetic nervous control, but THs induce the expression of a gene that encodes an enzyme called "uncoupling protein" (UCP), or "thermogenin." UCP is a mitochondrial protein that uncouples ATP synthesis and respiration, thereby greatly increasing heat production in BAT. There are more numerous TRs in BAT than in any other tissue except for the pituitary, and this tissue also contains the D2 deiodinase which generates locally much of the T_3 that binds to TR in BAT. Birds lack BAT, but recent investigations have demonstrated that THs induce UCP during muscle thermogenesis in birds (Raimbault et al. 2001; Collin, Buyse, et al. 2003; Collin, Taouis, et al. 2003).

Thyroid hormone effects on intermediary metabolism

THs play an important role in regulating intermediary metabolism, they stimulate fatty acid synthesis in the liver, and they act permissively to enhance catecholamine-induced lipolysis. It is generally believed that fatty acids constitute the principal energy source in hyperthyroid individuals. The TH-accelerated lipogenesis–lipolysis cycle seems paradoxical and has been termed the "futile cycling of fatty acids." However, it is energetically expensive in itself and may contribute up to 15% of the total heat production accounted for by TH action. THs also influence carbohydrate metabolism, but this probably contributes less to alterations in basal metabolic rate (BMR) than do the effects of fatty acid metabolism (Sestoft 1980). One way in which THs influence carbohydrate metabolism is by altering the expression of genes involved in the synthesis of glucose from carbohydrate, fat, and amino acid carbon backbones, a process called "gluconeogenesis." For example, TH alters expression of the gene encoding phosphoenolpyruvate carboxykinase (Hanson and Reshef 1997), a critical enzyme in the synthesis of glucose in muscle and liver.

Clinical disorders related to iodine deficiency and thyroid hormone synthesis

Deficient levels of circulating THs are generally referred to as "hypothyroidism." Hypothyroidism can be the result of genetic and epigenetic factors such as endocrine-disrupting contaminants. Congenital hypothyroidism results from defects in the development of normal thyroid tissue or the synthesis of TH, whereas endemic hypothyroidism results from deficiencies in the availability of dietary I^-. Both conditions can lead to retarded development of the nervous system. Endemic cretinism is a particularly severe form of mental retardation and occurs in areas where I^- availability is compromised during prenatal life. In many developed countries, congenital hypothyroidism is usually detected shortly after birth because of national postnatal monitoring of all newborns in many developed countries and is treated by exogenous TH administration whereas endemic hypothyroidism can be treated with dietary I^- supplementation. Endemic hypothyroidism is prevalent in underdeveloped countries with soil deficient in I^- and a lack of infrastructure to make iodized table salt or other iodine supplementation available on a wide scale (Meier and Burger 2000). So-called "iodine deficiency disorders" (IDDs) affect hundreds of millions of people in the world and are associated with goiter formation and CNS deficits such as deafness and cretinism (Toteja et al. 2004; Zimmerman et al. 2004). It is estimated that 750 million people suffer from goiter and that 43 million people have IDD-related brain damage and mental retardation worldwide (International Council for Control of Iodine Deficiency Disorder 2004). When adults develop IDDs, the effects can be reversed with I^- supplementation. However, the supplementation must be sustained because even short lapses in treatment can cause rapid deterioration of TH synthesis (Zimmerman et al. 2004).

An abnormal enlargement of the thyroid gland is generally referred to as a goiter. Enlargement of the thyroid occurs most commonly from iodine deficiency or from defects in the TH biosynthetic–secretion pathway that result in lower circulating T_4 levels. As a result of lower circulating TH, negative feedback to the pituitary gland leads to increased plasma TSH levels, which cause enlargement (hypertrophy) of the thyroid gland. Goiter also can be caused by exposure to goitrogenic chemicals such as PTU, which acts by inhibiting I^- organification, and perchlorate, which competitively inhibits I^- uptake by the thyroid. Hypothyroid goiters can also result from IDDs, as described above (Braverman and Utiger 2000). Hyperthyroid goiters occur less commonly but can be a symptom of Grave's disease. A special form of goiter may arise during pregnancy as a result of increased plasma TBG in the maternal circulation. This condition reduces circulating levels of "free" or unbound TH, thus leading to elevated TSH secretion and an enlarged thyroid that returns to normal size after delivery (Glinoer 1993).

Comparative thyroid physiology: generalized model

In all tetrapod vertebrates, thyrocytes are organized into follicles within an encapsulated gland or pair of glands and produce primarily T_4 and much smaller amounts of T_3. In most fishes, the follicles are not organized into a single gland but are widely distributed along the epithelial surface of the pharynx (Matsumoto and Ishii 1992). Exceptions to this scattered follicle distribution in fish include the Hawaiian parrotfish, in which the follicles are organized into 2 distinct lobes (Grau et al. 1986).

The biosynthetic pathway for THs is generally believed to be similar in all vertebrates. Biochemical and physiological evidence suggests that I^- uptake occurs via the activity of an NIS in all vertebrates, although no nonmammalian version of the protein has been fully cloned. Indirect evidence indicates that the mechanism of thyroidal I^- uptake is evolutionarily ancient, because potassium perchlorate, potassium thiocyanate and sodium perchlorate all reduce TH synthesis in the primitive lamprey (Manzon et al. 2001). Homologs of the TPO gene are expressed in the endostyle, a filter-feeding organ that sequesters I^- in urochordates, cephalochordates, and larval lampreys (Ogasawara et al. 1999).

THs are structurally identical throughout all the vertebrate classes. Because of this structural uniformity, the same antibodies to THs can be employed in radioimmunoassays for measuring THs in samples from all vertebrates. This is not the case for TSH, which is a glycoprotein. Although all vertebrates typically respond to heterologous forms of TSH with an increase in circulating concentrations of T_3 and T_4, the structure of TSH differs widely among vertebrate groups, and measurement of TSH requires species-specific antibodies. There are important differences between the vertebrate classes with respect to the details of thyroid gland function, TH-binding proteins, deiodinases, hypothalamic–pituitary control of thyroid function, and a number of other aspects of thyroid physiology and TH effects on target tissues. Some of these differences have been mentioned in this chapter, others will be covered in Chapters 4 through 7.

Mechanism of Perchlorate Action

Effects of perchlorate on thyroid gland function

It has been known since the 1950s and 1960s that perchlorate blocks I^- uptake by thyrocytes and some other cells that take up I^- (Stanbury and Wyngaarden 1952; Levens 1960; Lewitus et al. 1962). The potency with which this anion blocks I^- transport has made it a popular treatment for various clinical thyroid disorders and a useful experimental tool in some studies of basic thyroid function (Wolff 1998). A major advance in understanding the mechanism of perchlorate action came in 1996 with the cloning of the NIS by Carrasco and colleagues (Dai et al. 1996). Van Sande and colleagues (Van Sande et al. 2003) reinvestigated anion specificity in COS cells transiently transfected with human NIS cDNA and in FRTL-5 (Fisher rat thyroid cell line) cells. In this study, the investigators measured the direct uptake of $^{125}I^-$ as well as $^{186}ReO_4^-$ (as ^{186}Re sodium perrhenate) and $99mTcO_4^-$ (as 99mTc sodium pertechnetate). These latter 2 compounds possess oxyanions similar in size to perchlorate. Both $^{186}ReO_4^-$ and $99mTcO_4^-$ are used to image thyroid and GI tissues clinically. In all cases, the uptake (i.e., transport) of $^{125}I^-$ (and thereby, transport) was comparable to $^{186}ReO_4^-$ and $99mTcO_4^-$, with the rank-order potency for inhibition of uptake $ClO_4^- > ReO_4^- > I^- > SCN^- > ClO_3^- >> Br^-$. This relationship was found for both cell types and for all 3 radioligands, supporting many earlier studies indicating that perchlorate is indeed transported into thyrocytes by the NIS (Chow and Woodbury 1970; Clewell et al. 2004). These findings also resolve an earlier controversy regarding perchlorate transport by the NIS expressed in xenopus oocytes. These earlier studies revealed that the rank-order affinity of the NIS based on electrophysiological measurements in xenopus oocytes was $I^->ClO_3^->SCN^->Br^-$ (Eskandari et al. 1997). In these studies, investigators measured the changes in current across the oocyte membrane in the presence of 500 μM anion (Eskandari et al. 1997). Surprisingly, using the xenopus oocyte model, perchlorate and perrhenate were not transported (Eskandari et al. 1997), even though both anions are known to interact with the transporter and to potently block radiolabeled I^- uptake into oocytes expressing the NIS. Furthermore, these studies on the oocyte model contradicted earlier work showing that radiolabeled perchlorate is transported across the basal plasma membrane of thyrocytes and accumulates in colloid (Chow et al. 1969, 1970; Yu et al. 2002), and more recent studies showing that the kinetics of radiolabeled perrhenate and pertechnetate uptake in NIS-expressing tissues essentially mirrors that of I^- (Zuckier et al. 2004). Data that appear to explain these contradictions from different models were generated using FTRL-5 (a thyroid cell line) and Chinese (CHO) hamster ovary cells. In CHO cells stably expressing the rat NIS, perchlorate was a potent inhibitor of I^- transport but did not induce an inward current when bound to the transporter (Yoshida et al. 1997), meaning that it either was not transported across the plasma membrane or it was transported in a 1:1 stoichiometry with Na^+ ions, thereby eliminating any net current. In all of these studies, inward current was measured under the assumption that the NIS transports 2 Na^+ for every I^-. In FTRL-5 cells, a rat thyroid cell line that normally expresses the NIS, patch-clamping techniques revealed no net inward current across the plasma membrane when cells were exposed to perchlorate (Yoshida et al. 1998). Collectively, these findings illustrate that the expression model and the type methods used (direct or indirect measurements of perchlorate transport) can dramatically influence the analysis of perchlorate transport by the NIS.

Because perchlorate is transported by the NIS, extrathyroidal tissues that express this transporter would be expected to transport perchlorate. The NIS is expressed in epithelial cells from a number of tissues, including salivary glands, mammary tissues, stomach, and kidney (Lacroix et al. 2001; Spitzweg et al. 2001; Wapnir et al. 2003, also see De La Vieja et al. 2000 for a review). In kidney cells, the handling of I^- occurs much as it does in thyrocytes with the NIS moving I^- across the basolateral plasma membrane and pendrin chaperoning the ion across the apical cell surface (Gillam et al. 2004). In mammary tissue, the expression of the NIS is closely tied with the degree of lactation, being up-regulated during nursing and rapidly reduced at the cessation of nursing (Tazebay et al. 2000).

Whether environmentally relevant concentrations of perchlorate influence epithelial I^- transport in these extrathyroidal tissues is unknown.

Because perchlorate blocks the transport of I^- that is required to synthesize THs, exposure to this anion results in reduced TH synthesis and potential alterations in the thyroid homeostasis usually maintained by the HPT axis. Rats dosed with perchlorate via the drinking water, above a certain threshold, exhibit lowered plasma T_4 and T_3, elevated plasma TSH, histopathological correlates of altered thyroid function (colloid depletion and follicle cell hypertrophy), and increased thyroid weight (Siglin et al. 2000; York et al. 2003). It is important to note that some endpoints of thyroid status are more responsive to perchlorate than others; a comparison of the endpoints that have been used to evaluate thyroid status after perchlorate exposure is provided in Table 3-1. Not only do some endpoints respond differently than others, but there are important species- (Lewandoski et al. 2004) and age-related differences in the response of the HPT axis to perchlorate (see Chapters 4 through 7).

Perchlorate can be passed to offspring through lactation and can cross the placenta and thereby alter thyroid homeostasis in the fetus and in nursing neonates (Clewell et al. 2003; Mahle et al. 2003; Kirk et al. 2005). Oviparous vertebrates deposit perchlorate in their eggs, so these embryos also are exposed to perchlorate during development (see Chapters 5, 6, and 7).

Secondary effects of thyroid disruption

There is little indication that perchlorate has effects other than those on the thyroid gland. Thus, it appears that other systems will be little affected by perchlorate exposure except as a result of organismal hypothyroidism that occurs if the perchlorate exposure is sufficient to reduce circulating THs. It is expected that these downstream effects on target organs and tissues will be related to the severity of the hypothyroidism. Because perchlorate acts by creating an I^- deficiency within the thyroid, the large body of knowledge about iodine deficiency (section titled "Clinical disorders related to iodine defieciency and thyroid hormone synthesis" on page 60) has relevance to studies of perchlorate. As mentioned previously, THs are involved in the control of development in a number of body systems and tissues. The potential for permanent nervous system inadequacies as a result of hypothyroidism due to perchlorate exposure is of special concern because of the critical role of the nervous system in physiological control. However, inadequacies in the development and maturation of other systems, such as those affecting skeletal development and muscle

Table 3-1 Relative sensitivity of thyroid-related endpoints to perchlorate in various vertebrate groups

Endpoint	Group	Relative sensitivity[1]	Species and age	LOEC	Exposure duration	Reference
Direct thyroid effects						
Follicle cell height (hypertrophy)	Mammals	Moderate to high	Adult rats	$10\ mg \cdot kg^{-1} \cdot d^{-1}$	14 d, 90 d	Siglin et al. 2000
			Parents, F1 and F2 rats	$3\ mg \cdot kg^{-1} \cdot d^{-1}$	continuous	York et al. 2001a
			Adult rabbit does	$10\ mg \cdot kg^{-1} \cdot d^{-1}$	gestation	York et al. 2001b
			F1 rat pups	$3\ mg \cdot kg^{-1} \cdot d^{-1}$	gestation-lactation	York et al. 2004
	Amphibians	High	Larval *Xenopus laevis*	$59\ \mu g \cdot L^{-1}$	70 d	Goleman et al. 2002
			Larval *X. laevis*	$16\ \mu g \cdot L^{-1}$	14 d	Tietge et al. 2005
			Larval *Pseudacris triseriata*	$10\ mg \cdot L^{-1}$	unknown	Carr et al. 2003
			Adult *Acris crepitans*	$33\ \mu g \cdot L^{-1}$	unknown	USACE 2004
	Fishes	High	*Danio rerio*	$18\ mg \cdot L^{-1}$	8 wk	Patiño et al. 2003
			D. rerio	$1.13\ mg \cdot L^{-1}$	12 wk	Mukhi et al. 2005
Follicle cell number (hyperplasia)	Mammals	Low	Adult rats	$>10\ mg \cdot kg^{-1} \cdot d^{-1}$	14 d, 90 d	Siglin et al. 2000
			Parents, F1 rat pups	$3\ mg \cdot kg^{-1} \cdot d^{-1}$	continuous	York et al. 2001a
			Adult rabbit does	$>10\ mg \cdot kg^{-1} \cdot d^{-1}$	gestation	York et al. 2001b
			F1 rat pups	$3\ mg \cdot kg^{-1} \cdot d^{-1}$	gestation-lactation	York et al. 2004
	Amphibians	Low	Larval *X. laevis*	$>14\ mg \cdot L^{-1}$	70 d	Goleman et al. 2002
			Larval *P. triseriata*	$>10\ mg \cdot L^{-1}$	unknown	Carr et al. 2003
			Adult *A. crepitans*	$>33\ \mu g \cdot L^{-1}$	unknown	USACE 2004
	Fishes	Low	*D. rerio*	$18\ mg \cdot L^{-1}$	8 wk	Patiño et al. 2003
		No response	*D. rerio*	$>11.5\ mg \cdot L^{-1}$	12 wk	Mukhi et al. 2005
Colloid depletion	Mammals	Moderate to high	Adult rats	$10\ mg \cdot kg^{-1} \cdot d^{-1}$	14 d, 90 d	Siglin et al. 2000
		No response	Adult rabbit does	$>10\ mg \cdot kg^{-1} \cdot d^{-1}$	gestation	York et al. 2001b

Table 3-1, *cont'd.*

Endpoint	Group	Relative sensitivity[1]	Species and age	LOEC	Exposure duration	Reference
Direct thyroid effects, *cont'd*						
Colloid depletion, *cont'd.*	Amphibians	High	Larval *X. laevis*	16 $\mu g \cdot L^{-1}$	8 d	Tietge et al. 2005
			Larval P. *triseriata*	10 $mg \cdot L^{-1}$	unknown	Carr et al. 2003
		No response	Adult *A. crepitan*	>33 $\mu g \cdot L^{-1}$	unknown	USACE 2004
	Fishes	Moderate	*D. rerio*	18 $mg \cdot L^{-1}$	8 wk	Patiño et al. 2003
			D. rerio	11.5 $mg \cdot L^{-1}$	12 wk	Mukhi et al. 2005
Angiogenesis	Fishes	High	*D. rerio*	18 $mg \cdot L^{-1}$	8 wk	Patiño et al. 2003
			D. rerio	90 $\mu g \cdot L^{-1}$	12 wk	Mukhi et al. 2005
T_4 ring depletion	Fishes	Very high	*D. rerio*	11 $\mu g \cdot L^{-1}$	12 wk	Mukhi et al. 2005
Thyroid gland T_4	Birds	High	*Colinus virginianus*	50 $\mu g \cdot L^{-1}$ (~4 $\mu g \cdot kg^{-1} \cdot d^{-1}$)	2 wk	McNabb et al. 2004
Thyroid gland weight	Mammals	Variable	Adult rats	10 $mg \cdot kg^{-1} \cdot d^{-1}$	90 d	Siglin et al. 2000
			F1 adult rats	3 $mg \cdot kg^{-1} \cdot d^{-1}$	continuous	York et al. 2001a
			Parents, F1 rat pups	30 $mg \cdot kg^{-1} \cdot d^{-1}$	continuous	York et al. 2001a
		No response	Parents, F1 rat pups	>10 $mg \cdot kg^{-1} \cdot d^{-1}$	gestation	York et al. 2004
	Birds	Moderate	*C. virginianus*	500 $mg \cdot L^{-1}$ (~39 $mg \cdot kg^{-1} \cdot d^{-1}$)	2 wk	McNabb et al. 2004
Organismal thyroid activity						
Plasma T_4 (decreased)	Mammals	Low to high	Adult rats	0.01 $mg \cdot kg^{-1} \cdot d^{-1}$	90 d	Siglin et al. 2000
		No response[2]	F1 adult rats	0.3 $mg \cdot kg^{-1} \cdot d^{-1}$	continuous	York et al. 2001a
		No response	Parents, male rats	>30$mg \cdot kg^{-1} \cdot d^{-1}$	continuous	York et al. 2001a
			Parents, female rats	>30 $mg \cdot kg^{-1} \cdot d^{-1}$	continuous	York et al. 2001a
			Adult rabbit does	30 $mg \cdot kg^{-1} \cdot d^{-1}$	gestation	York et al. 2001b
			F1 rat pups	3 $mg \cdot kg^{-1} \cdot d^{-1}$	gestation	York et al. 2004

Table 3-1, *cont'd.*

Endpoint	Group	Relative sensitivity[1]	Species and age	LOEC	Exposure duration	Reference
	Birds	Low	*C. virginianus*	500 $mg \cdot L^{-1}$ (~39 $mg \cdot kg^{-1} \cdot d^{-1}$)	2 wk	McNabb et al. 2004
Plasma T_3 (decreased)	Mammals	Low to high	Adult rats	0.01 $mg \cdot kg^{-1} \cdot d^{-1}$	90 d	Siglin et al. 2000
			F1 adult rats	0.3 $mg \cdot kg^{-1} \cdot d^{-1}$	continuous	York et al. 2001a
		No response	Parents, female rats	>30$mg \cdot kg^{-1} \cdot d^{-1}$	continuous	York et al. 2001a
			Adult rabbit does	30 $mg \cdot kg^{-1} \cdot d^{-1}$	gestation	York et al. 2001b
			F1 rat pups	3 $mg \cdot kg^{-1} \cdot d^{-1}$	gestation	York et al. 2004
	Birds	Very low	*C. virginianus*	4000 $mg \cdot L^{-1}$ (~308 $mg \cdot kg^{-1} \cdot d^{-1}$)	2 wk	McNabb et al. 2004
Whole body T_4	Amphibians	Low	Larval *X. laevis*	14 $mg \cdot L^{-1}$	70 d	Goleman et al. 2002b
	Fishes	No response	*D. rerio*	>11.5 $mg \cdot L^{-1}$	12 wk	Patiño et al. 2003
Pituitary effect						
Plasma TSH	Mammals	Low to high	Adults rats	0.2 $mg \cdot kg^{-1} \cdot d^{-1}$	90 d	Siglin et al. 2000
			Male parents, F1 rat pups	30 $mg \cdot kg^{-1} \cdot d^{-1}$	continuous	York et al. 2001a
		No response	F1 rat pups	10 $mg \cdot kg^{-1} \cdot d^{-1}$	gestation	York et al. 2001a
		No response	Female parents	>30 $mg \cdot kg^{-1} \cdot d^{-1}$	continuous	York et al. 2001b
			Adult rabbit does	>100 $mg \cdot kg^{-1} \cdot d^{-1}$	gestation-lactation	

[1] Subjective evaluation relative to other endpoints used. Data on actual lowest observed effects concentrations for each endpoint are listed in Chapters 4 through 7.

[2] No response refers to a decrease in plasma T_3 concentrations; an anomalous increase in plasma T_3 occurred at 0.3 and 3.0 $mg \cdot kg^{-1} \cdot d^{-1}$, but not at 30 $mg \cdot kg^{-1} \cdot d^{-1}$ in adult female rats in this study

function, can seriously affect the survival and function of the animal and its ability to reproduce. Hypothyroidism may alter temperature regulation in homeotherms, and intermediary metabolism may be altered in all vertebrates. Specific studies investigating the influence of perchlorate on the organismal thyroid status and CNS, reproductive, and metabolic function in mammals are described in detail in Chapter 4.

Summary and Research Recommendations

The only known mechanism for perchlorate action is to competitively inhibit transepithelial I^- uptake by the NIS. The thyroid gland is an evolutionarily ancient endocrine organ that must accumulate I^- in order to produce the hormones T_4 and T_3, and is the major target for perchlorate action. In the thyroid gland, competitive inhibition of I^- transport by perchlorate results in reduced TH synthesis, which can be measured by reductions in plasma levels of T_4 and T_3. Reduced plasma levels of TH release pituitary thyrotropes from negative feedback, resulting in elevated plasma levels of TSH. The effects of elevated plasma TSH on the thyroid gland can be gauged by assessing the degree of follicle hypertrophy and hyperplasia and the degree of colloid depletion in the thyroid follicle.

The effects of perchlorate that have been reported on extrathyroidal tissues can be explained by the action of perchlorate at the thyroid gland and subsequent reductions TH synthesis. Given the wide spectrum of TH actions, reductions in circulating TH may negatively impact many aspects of growth, CNS development, and metabolic activity. Of particular concern are the potential effects of perchlorate-induced thyroid disruption in perinatal devolvement of the CNS because these effects may be long-lasting and have effects on learning and cognition that may be difficult to detect immediately after birth. Currently, the incidences of primary congenital hypothyroidism, as well as measurements of plasma TSH, are used to assess the effects of perchlorate exposure in newborn humans (Kelsh et al. 2003). The ability of PCH diagnosis to predict perchlorate-induced behavioral and learning deficits later in childhood is unknown.

Despite the overwhelming evidence that perchlorate binds to the NIS with greater affinity than I^-, there remains debate about the role of the NIS in transporting perchlorate across the basolateral plasma membrane of thyroid follicle cells. Moreover, the potential for perchlorate to interfere with apical I^- transport via the pendrin or hAIT proteins is not well studied. In addition, little work has focused on the consequences of perchlorate transport by extrathyroidal tissues. This lack of research seems to be an extraordinary oversight because the mechanisms responsible for transporting perchlorate into the bloodstream from dietary sources have not clearly been delineated in any vertebrate. The ability of perchlorate to interact with extrathyroidal NIS proteins, and the consequences of this interaction for whole animal I^- economy, should be the focus of future research.

References

Anderson GW, Schoonover CM, Jones SA. 2003. Control of thyroid hormone action in the developing rat brain. Thyroid 13:1039–1056.

Barter RA, Klaassen CD. 1992a. Rat liver microsomal UDP-glucuronosyltransferase activity toward thyroxine: characterization, induction, and form specificity. Toxicol Appl Pharmacol 115:261–267.

Barter RA, Klaassen CD. 1992b. UDP-glucuronosyltransferase inducers reduce thyroid hormone levels in rats by an extrathyroidal mechanism. Toxicol Appl Pharmacol 113:36–42.

Barter RA, Klaassen CD. 1994. Reduction of thyroid hormone levels and alteration of thyroid function by four representative UDP-glucuronosyltransferase inducers in rats. Toxicol Appl Pharmacol 128:9–17.

Bassett JH, Williams GR. 2003. The molecular actions of thyroid hormone in bone. Trends Endocrinol Metab 14:356-364.

Bernal J. 2002. Action of thyroid hormone in brain. J Endocrinol Investig 25:268–288.

Bianco AC, Salvatore D, Gereben B, Berry MJ, Larsen PR. 2002. Biochemistry, cellular and molecular biology, and physiological roles of the iodothyronine selenodeiodinases. Endocr Rev 23:38–89.

Bishop CM, McCabe CJ, Gittoes NJL, Butler PJ, Franklyn JA. 2000. Tissue-specific regulation of thyroid hormone receptor mRNA isoforms and target gene proteins in domestic ducks. J Endocrinol 165:607–615.

Book CB, Sun X, Ng YC. 1997. Developmental changes in regulation of the Na+, K+-ATPase a 3 isoform by thyroid hormone in ferret heart. Biochim Biophys Acta 1358:172–180.

Braverman LE, Utiger RD. 2000. Thyroid diseases: hypothyroidism. Part V. In: Braverman LE, Utiger RD, editors. Werner and Ingbar's the thyroid. 8th ed. Philadelphia (PA): Lippincott-Raven. p 719–860.

Breall JA, Rudolph AM, Heymann MA. 1984. Role of thyroid hormone in postnatal circulatory and metabolic adjustments. J Clin Investig 73:1418–1424.

Brouwer A, Morse DC, Lans MC, Schuur AG, Murk AJ, Klasson-Wehler E, Bergman A, Visser TJ. 1998. Interactions of persistent environmental organohalogens with the thyroid hormone system: mechanisms and possible consequences for animal and human health. In: Colborn T, vom Saal F, Short P, editors. Environmental endocrine-disrupting chemicals: neural, endocrine, and behavioral effects, advances in modern environmental toxicology, Vol. 24. Princeton (NJ): Princeton Scientific. p 69–98.

Canavan JP, Holt J, Goldspink DF. 1993. The role of thyroid hormones in the growth of striated muscles in the neonatal rat. Basic Appl Myol 3:85–92.

Carr JA, Urquidi LJ, Goleman WL, Hu F, Smith PN, Theodorakis CW. 2003. Ammonium perchlorate disruption of thyroid function in natural amphibian populations: assessment and potential impact. In: Linder G, editor. Multiple stressor effects in relation to declining amphibian populations. West Conshohocken (PA): American Soc of Testing and Materials. ASTM STP 1443. p 131–142.

Cassar-Malek I, Marchal S, Rochard P, Casas F, Wrutniak C, Samarut J, Cabello G. 1996. Induction of c-Erb A-AP-1 interactions and c-Erb A transcriptional activity in myoblasts by RXR: Consequences for muscle differentiation. J Biol Chem 271:11392–11399.

Chang L, Munro SL, Richardson SJ, Schreiber G 1999. Evolution of thyroid hormone binding by transthyretins in birds and mammals. Eur J Biochem 259:534–542.

Chow SY, Chang LR, Yen MS. 1969. A comparison between the uptakes of radioactive perchlorate and iodide by rat and guinea-pig thyroid glands. J Endocrinol 45:1–8.

Chow SY, Woodbury DM. 1970. Kinetics of distribution of radioactive perchlorate in rat and guinea-pig thyroid glands. J Endocrinol 47:207–218.

Clewell RA, Merrill EA, Narayanan L, Gearhart JM, Robinson PJ. 2004. Evidence for competitive inhibition of iodide uptake by perchlorate and translocation of perchlorate into the thyroid. Int J Toxicol 23:17–23.

Clewell RA, Merrill EA, Yu KO, Mahle DA, Sterner TR, Mattie DR, Robinson PJ, Fisher JW, Gearhart JM. 2003. Predicting fetal perchlorate dose and inhibition of iodide kinetics during gestation: a physiologically-based pharmacokinetic analysis of perchlorate and iodide kinetics in the rat. Toxicol Sci 73:235–255.

Collin A, Buyse J, Van As P, Darras VM, Malheiros RD, Moraes VMB, Reyns GE, Taouis M, Decuypere E. 2003. Cold-induced enhancement of avian uncoupling protein expression, heat production, and triiodothyronine concentrations in broiler chicks. Gen Comp Endocrinol 130:70–77.

Collin A, Taouis M, Buyse J, Ifuta N, Darras VM, Van As P, Malheiros RD, Moraes VMB, Decuypere E. 2003. Thyroid status, but not insulin status, affects expression of avian uncoupling protein mRNA in chicken. Am J Physiol Endocrinol Metab 284: E771–777.

Dai G, Levy O, Carrasco N. 1996. Cloning and characterization of the thyroid iodide transporter. Nature 379:458–460.

Danforth Jr E, Burger A. 1984. The role of thyroid hormones in the control of energy expenditure. Clin Endocrinol Metab 13:581–595.

Das B, Schreiber AM, Huang H, Brown DD. 2002. Multiple thyroid hormone–

induced muscle growth and death programs during metamorphosis in *Xenopus laevis*. Proc Natl Acad Sci U S A 99:12230–12235.

De La Vieja A, Dohan O, Levy O, Carrasco N. 2000. Molecular analysis of the sodium/iodide symporter: impact on thyroid and extrathyroid pathophysiology. Physiol Rev 80:1083–1105.

DiStefano III JJ. 1988. Excretion, metabolism and enterohepatic circulation pathways and their role in overall thyroid hormone regulation in the rat. Am Zool 28:373–387.

Dumitrescu AM, Liao XH, Best TB, Brockmann K, Refetoff S. 2004. A novel syndrome combining thyroid and neurological abnormalities is associated with mutations in a monocarboxylate transporter gene. Am J Hum Genet 74:168–175.

Dunn JT, Dunn AD. 2000. Thyroglobulin: chemistry, biosynthesis and proteolysis. In: Braverman LE, Utiger RD, editors. Werner and Ingbar's the thyroid, 8th ed. Philadelphia (PA): Lippincott-Raven. p 91–104.

Engler D, Burger A. 1984. The deiodination of the iodothyronines and their derivatives in man. Endocr Rev 5:151–184.

Eskandari S, Loo DD, Dai G, Levy O, Wright EM, Carrasco N. 1997. Thyroid Na+/I- symporter. Mechanism, stoichiometry, and specificity. J Biol Chem 272:27230–27238.

Everett LA, Glaser B, Beck JC, Idol JR, Buchs A, Heyman M, Adawi F, Hazani E, Nassir E, Baxevanis AD, Sheffield VC, Green ED. 1997. Pendred syndrome is caused by mutations in a putative sulphate transporter gene (PDS). Nat Genet 17:411–422.

Forhead AJ, Li J, Gilmour RS, Dauncey MJ, Fowden AL. 2002. Thyroid hormones and the mRNA of the GH receptor and IGFs in skeletal muscle of fetal sheep. Am J Physiol Endocrinol Metab 282:E80–E86.

Friesema EC, Ganguly S, Abdalla A, Manning Fox JE, Halestrap AP, Visser TJ. 2003. Identification of monocarboxylate transporter 8 as a specific thyroid hormone transporter. J Biol Chem 278:40128–40135.

Friesema EC, Jansen J, Milici C, Visser TJ. 2005. Thyroid hormone transporters. Vitam Horm 70:137–167.

Gillam MP, Sidhaye AR, Lee EJ, Rutishauser J, Stephan CW, Kopp P. 2004. Functional characterization of pendrin in a polarized cell system. Evidence for pendrin-mediated apical iodide efflux. J Biol Chem 279:13004–13010.

Glinoer D. 1993. Maternal thyroid function in pregnancy. Endocrinol Investig 16:374–378.

Goleman WL, Carr JA, Anderson TA. 2002. Environmentally relevant concentrations of ammonium perchlorate inhibit thyroid function and alter sex ratios in developing *Xenopus laevis*. Environ Toxicol Chem 21:590–597.

Grau EG, Helms LM, Shimoda SK, Ford CA, LeGrand J, Yamauchi K. 1986. The thyroid gland of the Hawaiian parrotfish and its use as an in vitro model system. Gen Comp Endocrinol 61:100–108.

Haddow JE, Palomaki GE, Allan WC, Williams JR, Knight GJ, Gagnon J, O'Heir CE, Mitchell ML, Hermos RJ, Waisbren SE, Faix JD, Klein RZ. 1999. Maternal thyroid deficiency during pregnancy and subsequent neuropsychological development of the child. N Engl J Med 341:549–555.

Hanson RW, Reshef L. 1997. Regulation of phosphoenolpyruvate carboxykinase (GTP) gene expression. Ann Rev Biochem 66:581–611.

Hennemann G, Docter R, Friesema EC, de Jong M, Krenning EP, Visser TJ. 2001. Plasma membrane transport of thyroid hormones and its role in thyroid hormone metabolism and bioavailability. Endocr Rev 22:451–76.

Heuer H, Maier MK, Iden S, Mittag J, Friesema EC, Visser TJ, Bauer K. 2005. The monocarboxylate transporter 8 linked to human psychomotor retardation is highly expressed in thyroid hormone-sensitive neuron populations. Endocrinology 146:1698–700.

Howdeshell KL. 2002. A model of the development of the brain as a construct of the thyroid system. Environ Health Perspect 110 Suppl 3:337–348.

International Council for Control of Iodine Deficiency Disorder. 2004. Available from: http://www.iccidd.org. Accessed 14 Jun 2004.

Kelsh MA, Buffler PA, Daaboul JJ, Rutherford GW, Lau EC, Barnard JC, Exuzides AK, Madl AK, Palmer LG, Lorey FW. 2003. Primary congenital hypothyroidism, newborn thyroid function, and environmental perchlorate exposure among residents of a Southern California community. J Occup Environ Med 45:1116–1127.

Kirk AB, Martinelango PK, Tian K, Dutta A, Smith EE, Dasgupta PK. 2005. Perchlorate and iodide in dairy and breast milk. Environ Sci Technol 39:2011–2017.

Lacroix L, Mian C, Caillou B, Talbot M, Filetti S, Schlumberger M, Bidart JM. 2001. Na(+)/I(-) symporter and Pendred syndrome gene and protein expressions in human extra-thyroidal tissues. Eur J Endocrinol 144:297–302.

Larsson M, Pettersson T, Carlstrom A. 1985. Thyroid hormone binding in serum of 15 vertebrate species: isolation of thyroxine-binding globulin and prealbumin analogs. Gen Comp Endocrinol 58:360–375.

Legrand J. 1986. Thyroid hormone effects on growth and development. In: Hennemann G, editor. Thyroid hormone metabolism. New York (NY): Marcel Dekker. p 503–534.

Leonard JL, Koehrle J. 2000. Intracellular pathways of iodothyronine metabolism. In: Braverman LE, Utiger RD, editors. In: Werner and Ingbar's the thyroid, 8th ed. Philadelphia (PA): Lippincott-Raven. p 136–173.

Levens L. 1960. Biological action of perchlorates. In: Schumacher JC, editor. Perchlorates: their properties, manufacture and uses. New York (NY): Reinhold. p 168–186.

Lewandowski TA, Seeley MR, Beck BD. 2004. Interspecies differences in susceptibility to perturbation of thyroid homeostasis: a case study with perchlorate. Regul Toxicol Pharmacol 39:348–362.

Lewitus Z, Guttmann S, Anbar M. 1962. Effect of thyroid-stimulating hormone (TSH) on the accumulation of perchlorate and fluoroborate ions in the thyroid glands of rats. Endocrinology 70:295–297.

Lisi S, Pinchera A, McCluskey RT, Willnow TE, Refetoff S, Marcocci C, Vitti P, Menconi F, Grasso L, Luchetti F, Collins AB, Marino M. 2003. Preferential megalin-mediated transcytosis of low-hormonogenic thyroglobulin: a control mechanism for thyroid hormone release. Proc Natl Acad Sci U S A 100:14858–14863.

Mahle DA, Yu KO, Narayanan L, Mattie DR, Fisher JW. 2003. Changes in cross-fostered Sprague-Dawley rat litters exposed to perchlorate. Int J Toxicol 22:87–94.

Manzon RG, Holmes JA, Youson JH. 2001. Variable effects of goitrogens in inducing precocious metamorphosis in sea lampreys (*Petromyzon marinus*). J Exp Zool 289:290–303.

Marians RC, Ng L, Blair HC, Unger P, Graves PN, Davies TF. 2002. Defining thyrotropin-dependent and -independent steps of thyroid hormone synthesis by using thyrotropin receptor-null mice. Proc Natl Acad Sci U S A 99:15776–15781.

Marino M, Pinchera A, McCluskey RT, Chiovato L. 2001. Megalin in thyroid physiology and pathology. Thyroid 11:47–56.

Matsumoto A, Ishii S, editors. 1992. Atlas of endocrine organs, New York (NY): Springer-Verlag. p 307.

McNabb FMA, Jang DA, Larsen CT. 2004. Does thyroid function in developing birds adapt to sustained ammonium perchlorate exposure? Toxicol Sci 82:106–113.

Meier CA, Burger AG. 2000. Effects of drugs and other substances on thyroid hormone synthesis and metabolism. In: Braverman LE, Utiger RD, editors. Werner and Ingbar's the thyroid, 8th ed. Philadelphia (PA): Lippincott-Raven. p 265–280.

Merkulova T, Keller A, Oliviero P, Marotte F, Samuel JL, Rappaport L, Lamande N, Lucas M. 2000. Thyroid hormones differentially modulate enolase isozymes during rat skeletal and cardiac muscle development. Am J Physiol 278:E330–E339.

Mills SA, Savage DD. 1988. Evidence of hypothyroidism in the genetically epilepsy-prone rat. Epilepsy Res 2:102–110.

Moussavi R, Meisami E, Timiras S. 1985. Compensation, cell proliferation and growth in the rat heart after postnatal hypothyroidism. Am J. Physiol 248: E381–387.

Mukhi S, Carr JA, Anderson TA, Patino R. 2005. Development and validation of new biomarkers of perchlorate exposure in fishes. Environ Toxicol Chem 24:1107–1115.

Nilsson M. 2001. Iodide handling by the thyroid epithelial cell. Exp Clin

Endocrinol Diabetes 2001:13–17.

Ogasawara M, Di Lauro R, Satoh N. 1999. Ascidian homologs of mammalian thyroid peroxidase genes are expressed in the thyroid-equivalent region of the endostyle. J Exp Zool 285:158–169.

Ohmori M, Endo T, Harii N, Inaya T. 1998. A novel thyroid transcription factor is essential for thyrotropin-induced up-regulation of Na+/I- symporter gene expression. Mol Endocrinol 12:727–736.

Palha JA, Fernandes R, de Escobar GM, Episkopou V, Gottesman M, Saraiva MJ. 2000. Transthyretin regulates thyroid hormone levels in the choroid plexus, but not in the brain parenchyma: study in a transthyretin-null mouse model. Endocrinology 141:3267–3272.

Patiño R, Wainscott MR, Cruz-Li EI, Balakrishnan S, McMurry C, Blazer VS, Anderson TA. 2003. Effects of ammonium perchlorate on the reproductive performance and thyroid follicle histology of zebrafish. Environ Toxicol Chem 22:1115–1121.

Pizzagalli F, Hagenbuch B, Stieger B, Klenk U, Folkers G, Meier PJ. 2002. Identification of a novel human organic anion transporting polypeptide as a high affinity thyroxine transporter. Mol Endocrinol 16:2283–2296.

Pop VJ, Kuijpens JL, van Baar AL, Verkerk G, van Son MM, de Vijlder JJ, Vulsma T, Wiersings WM, Drexhage HA, Vader JL. 1999. Low maternal free thyroxin concentrations during early pregnancy are associated with impaired psychomotor development in infancy. Clin Endocrinol 50:149–155.

Porterfield SP, Hendrich CE. 1993. The role of thyroid hormones in prenatal and neonatal neurological development: current perspectives. Endocr Rev 14:94–106.

Prapunpoj P, Richardson SJ, Schreiber G. 2002. Crocodile transthyretin: structure, function, and evolution. Am J Physiol Regul Integr Comp Physiol 283:R885–896.

Raimbault S, Dridi S, Denjean F, Lachuer J, Couplan E, Bouillard F, Bordas A, Duchamp C, Taouis M, Ricquier D. 2001. An uncoupling protein homologue putatively involved in facultative thermogenesis in birds. Biochem J 353:441–444.

Robbins J. 1996. Thyroid hormone transport proteins and the physiology of hormone binding. In: Braverman LE, Utiger RD, editors. Werner and Ingbar's the thyroid, 7th ed. Philadelphia (PA): Lippincott-Raven. p 96–110.

Rodriguez AM, Perron B, Lacroix L, Caillou B, Leblanc G, Schlumberger M, Bidart JM, Pourcher T. 2002. Identification and characterization of a putative human iodide transporter located at the apical membrane of thyrocytes. J Clin Endocrinol Metab 87:3500–3503.

Royaux IE, Suzuki K, Mori A, Katoh R, Everett LA, Kohn LD, Green ED. 2000. Pendrin, the protein encoded by the Pendred syndrome gene (PDS), is an apical porter of iodide in the thyroid and is regulated by thyroglobulin in FRTL-5 cells. Endocrinology 141:839–845.

Scanes CG, Campbell R, Harvey S, King DB, Malamed S, Perez F. 1986. Growth hormone in birds: a comparative perspective. In: Gaillard PJ, Boer HH, editors. Comparative endocrinology: developments and directions. New York (NY): Alan R. Liss. p 115–136.

Schreiber G. 2002. Beyond carrier proteins. The evolutionary and integrative roles of transthyretin in thyroid hormone homeostasis. J Endocrinol 175:61–73.

Sestoft L. 1980. Metabolic aspects of the calorigenic effect of thyroid hormone in mammals. Clin Endocrinol 13:489–506.

Siglin JC, Mattie DR, Dodd DE, Hildebrandt PK, Baker WH. 2000. A 90-day drinking water toxicity study in rats of the environmental contaminant ammonium perchlorate. Toxicol Sci 57:61–74.

Silva JE, Dick TE, Larsen PR. 1978. The contribution of local tissue thyroxine monodeiodination to the nuclear 3,5,3'-triiodothyronine in pituitary, liver, and kidney of euthyroid rats. Endocrinology 103:1196–1207.

Spitzweg C, Dutton CM, Castro MR, Bergert ER, Goellner JR, Heufelder AE, Morris JC. 2001. Expression of the sodium iodide symporter in human kidney. Kidney Int 59:1013–1023.

Stanbury JB, Wyngaarden JB. 1952. Effect of perchlorate on the human thyroid gland. Metabolism 1:533–539.

Tazebay UH, Wapnir IL, Levy O, Dohan O, Zuckier LS, Zhao QH, Deng HF, Amenta PS, Fineberg S, Pestell RG, Carrasco N. 2000. The mammary gland iodide transporter is expressed during lactation and in breast cancer. Nat Med 6:871–878.

Tietge JE, Holcombe GW, Flynn KM, Kosian PA, Korte JJ, Anderson LE, Wolf DC, Degitz SJ. 2005. Metamorphic inhibition of *Xenopus laevis* by sodium perchlorate: effects on development and thyroid histology. Environ Toxicol Chem 24:926–933.

Toteja GS, Singh P, Dhillon BS, Saxena BN. 2004. Iodine deficiency disorders in 15 districts of India. Indian J Pediatr 71:25–28.

[USACE] United States Army Corps of Engineers. 2004. Final Report: Bosque and Leon River Watersheds Study. Available from: http://www.swf.usace.army.mil/pubdata/ppmd/Perchlorate/deliverable final.html. Accessed 7/04.

[USEPA] United States Environmental Protection Agency. 1997. Special report on environmental endocrine disruption: an effects assessment and analysis. February 1997. USEPA Office of Research and Development. EPA/630/R-96/012. 111 p. Available from: http://www.epa.gov/ORD/WebPubs/endocrine/. Accessed 7/04.

Van Sande J, Massart C, Beauwens R, Schoutens A, Costagliola S, Dumont JE, Wolff J. 2003. Anion selectivity by the sodium iodide symporter. Endocrinology 144:247–252.

Visser TJ, Kaptein E, Glatt H, Bartsch I, Hagen M, Coughtrie MWH. 1998. Characterization of thyroid hormone sulfotransferases. Chem Biol Interact

109:279–291.

Wapnir IL, van de Rijn M, Nowels K, Amenta PS, Walton K, Montgomery K, Greco RS, Dohan O, Carrasco N. 2003. Immunohistochemical profile of the sodium/iodide symporter in thyroid, breast, and other carcinomas using high density tissue microarrays and conventional sections. J Clin Endocrinol Metab 88:1880–1888.

Wickenden AD, Kaprielian R, Parker TG, Jones OT, Backx PH. 1997. Effect of development and thyroid hormone on K+ currents and K+ channel gene expression in rat ventricle. J Physiol 504:271–286.

Wolff J. 1998. Perchlorate and the thyroid gland. Pharmacol Rev 50:89–105.

Wrutniak-Cabello C, Casas F, Cabello G. 2001. Thyroid hormone action in mitochondria. J Mol Endocrinol 26:67–77.

Yamauchi K, Takeuchi H, Overall M, Dziadek M, Munro SLA, Schreiber G. 1993. Structural characteristics of bullfrog (Rana catesbiana) transthyretin and its cDNA: comparison of its pattern of expression during metamorphosis with that of lipocalin. Eur J Biochem 256:287–296.

York RG, Barnett Jr J, Brown WR, Garman RH, Mattie DR, Dodd D. 2004. A rat neurodevelopmental evaluation of offspring, including evaluation of adult and neonatal thyroid, from mothers treated with ammonium perchlorate in drinking water. Int J Toxicol 23:191–214.

York RG, Brown WR, Girard MF, Dollarhide JS. 2001a. Two-generation reproduction study of ammonium perchlorate in drinking water in rats evaluates thyroid toxicity. Int J Toxicol 20:183–197.

York RG, Brown WR, Girard MF, Dollarhide JS. 2001b. Oral (drinking water) developmental toxicity study of ammonium perchlorate in New Zealand White rabbits. Int J Toxicol 20:199–205.

York RG, Funk KA, Girard MF, Mattie D, Strawson JE. 2003. Oral (drinking water) developmental toxicity study of ammonium perchlorate in Sprague-Dawley rats. Int J Toxicol 22:453–464.

Yoshida A, Hisatome I, Taniguchi S, Sasaki N, Yamamoto Y, Miake J, Fukui H, Shimizu H, Okamura T, Okura T, Igawa O, Shigemasa C, Green ED, Kohn LD, Suzuki K. 2004. Mechanism of iodide/chloride exchange by Pendrin. Endocrinology 145:4301–4308.

Yoshida A, Sasaki N, Mori A, Taniguchi S, Mitani Y, Ueta Y, Hattori K, Sato R, Hisatome I, Mori T, Shigemasa C, Kosugi S. 1997. Different electrophysiological character of I-, ClO4-, and SCN- in the transport by Na+/I- symporter. Biochem Biophys Res Commun 231:731–734.

Yoshida A, Sasaki N, Mori A, Taniguchi S, Ueta Y, Hattori K, Tanaka Y, Igawa O, Tsuboi M, Sugawa H, Sato R, Hisatome I, Shigemasa C, Grollman EF, Kosugi S. 1998. Differences in the electrophysiological response to I- and the inhibitory anions SCN^- and $ClO^-{}_4$, studied in FRTL-5 cells. Biochim Biophys Acta 1414:231–237.

Yu KO, Narayanan L, Mattie DR, Godfrey RJ, Todd PN, Sterner TR, Mahle

DA, Lumpkin MH, Fisher JW. 2002. The pharmacokinetics of perchlorate and its effect on the hypothalamus-pituitary-thyroid axis in the male rat. Toxicol Appl Pharmacol 182:148–159.

Zhang J, Lazar MA. 2000. The mechanism of action of thyroid hormones. Ann Rev Physiol 62:439–466.

Zimmermann MB, Wegmuller R, Zeder C, Torresani T, Chaouki N. 2004. Rapid relapse of thyroid dysfunction and goiter in school-age children after discontinuation of salt iodization. Am J Clin Nutr 79:642–645.

Zoeller RT, Dowling ALS, Herzig CTA, Iannacone EA, Gauger KJ, Bansal R. 2002. Thyroid hormone, brain development, and the environment. Environ Health Perspect 110:355–361.

Zuckier LS, Dohan O, Li Y, Chang CJ, Carrasco N, Dadachova E. 2004. Kinetics of perrhenate uptake and comparative biodistribution of perrhenate, pertechnetate, and iodide by NaI symporter-expressing tissues in vivo. J Nucl Med 45:500–507.

four

Perchlorate Effects on Mammals

Ernest E Smith, FM Anne McNabb, John Isanhart, Philip N Smith, Kenneth R Dixon

General Considerations

Risks posed to the development and subsequent health of animals exposed to environmental endocrine-disrupting contaminants such as perchlorate have not been thoroughly researched. It is generally accepted that exposure to perchlorate affects the thyroid gland and thyroid hormone (TH) concentrations in adult and developing animals. Perchlorate-induced hypothyroidism could potentially lead to secondary effects in target organs that are ultimately manifested as neurological, developmental, reproductive, or metabolic changes. This chapter provides an overview of studies on perchlorate in mammalian models—both laboratory and field animals—especially its effects on thyroid function.

Perchlorate Exposure in Adult and Developing Mammals

Under laboratory and field conditions, exposure of animals to perchlorate can occur through ingestion of various food sources, soil, and drinking water. Perchlorate is readily absorbed from the gastrointestinal tract and is excreted rapidly and virtually unchanged in the urine (Wolff 1998). There is no evidence that perchlorate is metabolized to any significant extent (Pontius et al. 2000).

Perchlorate Ecotoxicology. Ronald J Kendall and Philip N Smith, editors.

ISBN-10: 1-880611-90-2 ISBN-13: 978-1-880611-90-6

Laboratory and field exposures

For laboratory studies, some publications have presented only the concentration of perchlorate in the drinking water that test animals consumed, others have measured drinking rates to determine actual perchlorate intake, and at least 1 study measured tissue perchlorate concentrations in mammals in relation to perchlorate exposure. Thuett et al. (2002b) found that hepatic perchlorate ion concentrations in deer mice (*Peromyscus maniculatus*) exposed to 117.4 $mg \cdot L^{-1}$ ammonium perchlorate (AP) in drinking water ranged from 1.0 to 2.5 mg perchlorate $ion \cdot kg^{-1}$ dw.

Exposure of wildlife to perchlorate in the field is most often associated with ground and surface water as well as food sources such as plant material (grasses, seeds, and forbs) and associated soil (Smith et al. 2004). Mammals living in contaminated aquatic ecosystems and at terrestrial–aquatic interfaces have the highest likelihood of being exposed to perchlorate and having subsequent adverse responses. Terrestrial mammals at risk include herbivorous and granivorous rodents, livestock, and some omnivorous species inhabiting contaminated riparian areas.

Tissue analyses of wild mammals collected from contaminated areas have identified elevated concentrations of perchlorate ion in liver and kidney tissues (Smith et al. 2001). A preliminary assessment of perchlorate concentrations in environmental compartments and tissues of various wildlife species at the Longhorn Army Ammunition Plant (LHAAP), Karnack, Texas (see Chapter 2 for details), resulted in the detection of perchlorate in water, sediment, and soil samples from all the sites sampled at LHAAP. Perchlorate was detected in the tissues of several rodent species from the LHAAP site (Smith et al. 2001). A composite sample of livers from cotton mice (*Peromyscus gossypinus*) captured near a groundwater discharge area known as Harrison Bayou contained perchlorate at 356 $ng \cdot g^{-1}$ dw; however, additional composite liver samples from cotton mice and harvest mice (*Reithrodontomys fulvescens*) caught in the same area contained no detectable perchlorate. Two composite liver samples from harvest mice collected from an area within 50 m of a perchlorate-grinding facility contained 1120 and 2328 ng $perchlorate \cdot g^{-1}$ dw. Seeds and invertebrates make up the major portion of the diet of harvest mice, but the mice also consume grasses and herbs. Thus, it was surprising that neither of 2 samples collected at this facility from herbivorous mammals, a cotton mouse composite liver sample and a single Southern short-tailed shrew (*Blarina carolinensis*) sample, contained any detectable perchlorate. Composite samples consisting of thyroid tissue from harvest mice, cotton mice, and cotton rats captured at the Harrison Bayou discharge area and other sites within the LHAAP had perchlorate concentrations ranging from 0.59 to

2.17 $\mu g \cdot g^{-1}$ dw. In a separate study, Parsons (2001), using an analytical method with a lower sensitivity limit of 300 to 400 μg perchlorate $\cdot kg^{-1}$, detected perchlorate in mammals from areas where concentrations in surface soils were high (i.e., $\geq 9000\ \mu g \cdot kg^{-1}$).

In general, most rodents do not drink large volumes of water. High concentrations of perchlorate in seeds of plants near the perchlorate-grinding facility at the LHAAP could have contributed significantly to exposure among granivorous mammalian species, particularly rodents. Some rodents consume foods with relatively high preformed water content and therefore have moderate water needs. Desert rodents and those from arid areas have very low water needs because they have evolved mechanisms that minimize water losses (Vaughn 1986). In rodents with low to moderate water needs, food and soil perchlorate concentrations might therefore be expected to correlate more closely than water with rodent tissue perchlorate concentrations. Soil can make up a significant portion of the diet of wild rodents (Beyer et al. 1994) and thus may be a significant source or a more constant source of perchlorate exposure in rodents than is vegetative matter. Therefore, soil perchlorate concentration may be important as a component in assessing mammalian perchlorate exposures.

Clearly, mammalian wildlife inhabiting contaminated aquatic–terrestrial habitats can be exposed to perchlorate through ingestion of contaminated water, food, and soil. But many factors (both physiological and environmental) can result in transient alterations in tissue perchlorate concentrations and may be partly responsible for the varying concentrations of perchlorate that have been detected in wildlife tissue. For example, perchlorate is excreted rapidly, so if intake is short-lived, there could be transient high levels followed by rapid decreases in tissue perchlorate concentrations. Smith et al. (2005) investigated perchlorate exposure in raccoons inhabiting the LHAAP site. Raccoons consume approximately 0.083 g water $\cdot g^{-1}$ body weight $\cdot d^{-1}$ along with a diet principally composed of plant and animal matter (insects, amphibians, crayfish, fish, and rodents). The diet of raccoons may also include a significant amount of soil from feeding and grooming. Based on the potential for exposure through water ingestion, diet, and soil consumption, raccoons were monitored for perchlorate exposure by examining plasma perchlorate concentrations. However, few of the raccoons sampled had detectable plasma perchlorate concentrations. The authors suggest that, despite the fact that the raccoons live in a perchlorate-contaminated area, there may be marked temporal and spatial variations in actual perchlorate exposure, and rapid urinary excretion of perchlorate may prevent sustained perchlorate retention in the body.

Herbivorous livestock may be exposed to perchlorate through food and water. Cheng et al. (2004) reported that perchlorate was present in 2 of 7 biweekly plasma samples from 1 of 2 heifer calves raised on a pasture impacted by perchlorate (25 $\mu g \cdot L^{-1}$ in drinking water), but perchlorate was not detected in any samples from the control heifers during the 14-week study (Figure 4-1). The perchlorate exposure in this study had no effect on circulating concentrations of T_4 or T_3 (Figures 4-2a and 4-2b). Perchlorate was not detected in any of the tissues, typically used as human food sources, from heifers raised on the contaminated pasture.

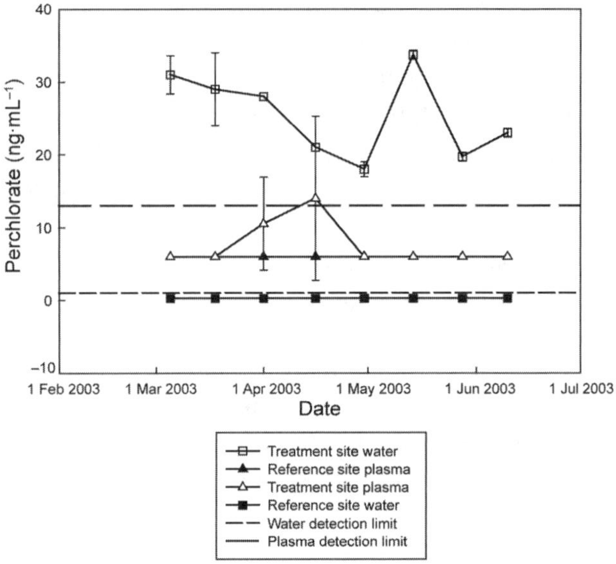

Figure 4-1 Relationship between perchlorate exposure in drinking water and perchlorate in heifer plasma. Each data point is the mean ± SD of duplicate measurements. For graphing and calculation purposes, non-detects were given a value of ½ the respective detection limit for perchlorate in water and plasma. (Reprinted with permission from Cheng et al. 2004. Copyright 2004 American Chemical Society)

Prenatal and early postnatal exposure unique to mammals

Exposure to perchlorate during development can potentially influence or modify subsequent developmental events, but information on the specific biological mechanisms whereby these processes may be altered is lacking (Siglin et al. 2000). Based on the levels of perchlorate present in various ecosystem compartments (see Chapter 2 for details) and the identification of perchlorate in commercially available cow's milk (Kirk et al. 2003) and human breast milk (Kirk et al. 2005), it is apparent that perchlorate exposure occurs through milk to nursing offspring in mammals. It is therefore important to evaluate maternal exposure and the rate of perchlorate transfer to fetuses and neonates and to assess the effects of perchlorate on thyroid function at these stages of development. Except for the recent study by Kirk et al. (2005), there are very little data on maternal milk transfer,

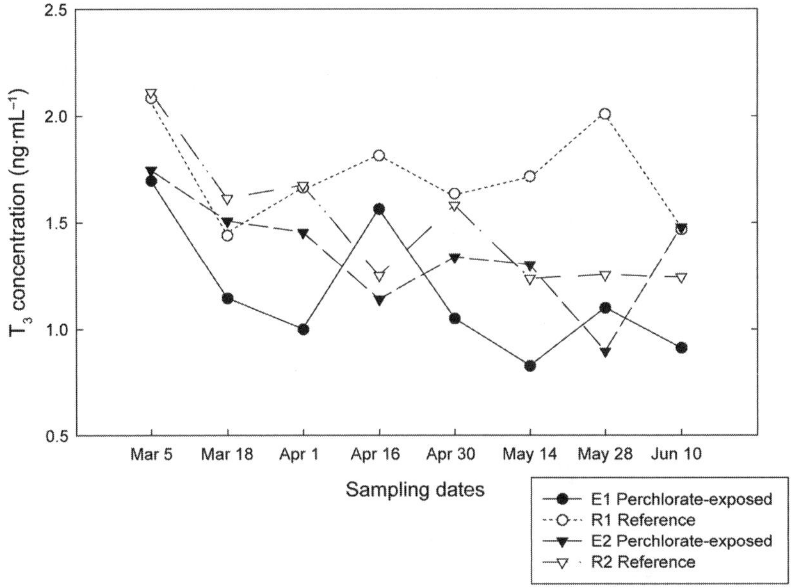

Figure 4-2a Total T_3 concentrations in plasma from 4 heifer calves on either perchlorate-contaminated (E1, E2) or reference pastures (R1, R2) near McGregor, TX 2003 (Reprinted with permission from Cheng et al. 2004. Copyright 2004 American Chemical Society)

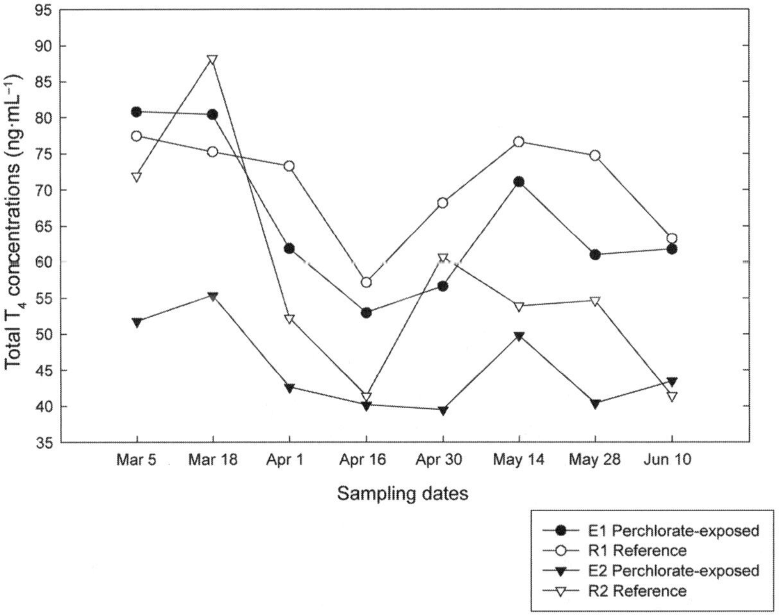

Figure 4-2b Total T_4 concentrations in plasma from 4 heifer calves on either perchlorate-contaminated (E1, E2) or reference pastures (R1, R2) near McGregor, TX 2003 (Reprinted with permission from Cheng et al. 2004. Copyright 2004 American Chemical Society)

despite its clear importance. In general, there is little information available on maternal transfer to developing offspring during gestation and lactation, in the dialogue about "sensitive subpopulations" in the clinical literature.

Perchlorate Effects on Thyroid Function

Studies of perchlorate effects on thyroid function in laboratory mammals

Uptake of radioactive perchlorate and iodide, as well as the kinetics of distribution of radioactive perchlorate in rat and guinea pig thyroid glands, has been demonstrated (Chow et al. 1969; Chow and Woodbury 1970). Morphological changes of the thyroid gland are associated with exposure to perchlorate in both wildlife and laboratory animals. Perchlorate administered to Sprague-Dawley rats in a subchronic toxicity test for 14 or 90 d in drinking water at dosages ranging from 0.01 to 10 $mg \cdot kg^{-1} \cdot d^{-1}$ AP salt significantly increased thyroid weight and induced follicular cell hypertrophy with microfollicle and colloid depletion only in the 10 $mg \cdot kg^{-1} \cdot d^{-1}$ group (Siglin et al. 2000). Following chronic exposure to perchlorate (1%) for up to 730 d, Wistar rats displayed increased thyroid weight and some follicular hyperplasia (Kessler and Kruskemper 1966). In this study, 4 of 11 treated rats developed benign thyroid tumors after 2 years. Sustained thyroid-stimulating hormone (TSH) exposure is associated with an increased incidence of thyroid malignancy (Pisarev and Kleiman de Pisarev 1980; Capen 1994).

Thyroidal tissue mass changes have been observed in the parent animals exposed to perchlorate during developmental toxicity experimentation. Exposure to AP (0.0 to 30 $mg \cdot kg^{-1} \cdot d^{-1}$), from 14 d before cohabitation to gestation day 21, resulted in increased maternal thyroid weight (both absolute and relative) and histopathology in the 30 $mg \cdot kg^{-1} \cdot d^{-1}$ treatment group (York et al. 2003). In a separate study, York et al. (2001a) evaluated the developmental toxicity of perchlorate salts (0.0 to100 $mg \cdot kg^{-1} \cdot d^{-1}$) in New Zealand White rabbits and observed maternal thyroid hypertrophy at $\geq$ 10 $mg \cdot kg^{-1} \cdot d^{-1}$ doses. Based on the effects of perchlorate in the thyroid gland in both studies, the maternal no-observable-adverse-effect level (NOAEL) was 1.0 $mg \cdot kg^{-1} \cdot d^{-1}$ for rats and rabbits. In contrast, the NOAEL for F1 adult rats exposed to perchlorate-treated water of the same concentration as was used for exposing their parents (up to 30.0 $mg \cdot kg^{-1} \cdot d^{-1}$) was considered to be 0.3 $mg \cdot kg^{-1} \cdot d^{-1}$, while the lowest-observable-adverse-effect level was reported to be 3.0 $mg \cdot kg^{-1} \cdot d^{-1}$ (York et al. 2001b).

The effects of AP exposure on rat pups during gestation and lactation have been evaluated by York et al. (2004). At postnatal day 5, pups from treated mothers (0.1 and 3.0 $mg \cdot kg^{-1} \cdot d^{-1}$ AP doses to the mothers) had pronounced colloid depletion and thyroid hypertrophy compared to those from control mothers. Thyroid hyperplasia also was documented in the pups of the 3.0 $mg \cdot kg^{-1} \cdot d^{-1}$ maternal treatment group. These effects on the pups appeared to be reversible, at maternal doses of 1.0 $mg \cdot kg^{-1} \cdot d^{-1}$ and higher; at postnatal day 59, there was only slight and nonsignificant colloid depletion and thyroid hypertrophy compared to controls. However, in an earlier study by these authors, F1 generation adult rats from a 2-generation reproductive study with continuous maternal exposure to perchlorate (0.0 to 30 $mg \cdot kg^{-1} \cdot d^{-1}$) had significantly increased relative thyroid weights for all dose groups for females and in the 3.0 and 30 $mg \cdot kg^{-1} \cdot d^{-1}$ dose groups for males (York et al. 2001b). Perchlorate-induced thyroid hypertrophy, and the incidence and severity of hyperplasia, increased in a dose-related manner. The F2 generation had significant increases in these variables, for both males and females, only at the highest maternal exposure (30 $mg \cdot kg^{-1} \cdot d^{-1}$).

The regulation of thyroid homeostasis in mammals is complex (York et al. 2004). Perchlorate exposure decreases iodide uptake, which in turn disrupts the biosynthesis and secretion of THs. The resultant decrease in circulating THs results in long-term activation of the hypothalamic–pituitary–thyroid (HPT) axis, resulting in chronic increases in TSH release, which in turn causes histological and physiological changes in the thyroid gland of mammals. Some studies of laboratory rats have found increased serum TSH with perchlorate exposure. Siglin et al. (1998) found increased circulating TSH concentrations at AP exposures of ≥0.2 $mg \cdot kg^{-1} \cdot d^{-1}$ in male and at 0.05 $mg \cdot kg^{-1} \cdot d^{-1}$ in female rats. These data suggested that increased TSH was a very sensitive indicator of altered thyroid function with perchlorate exposure. In contrast, the studies of York et al. (2001a, 2001b, 2004) found TSH to be significantly increased compared to controls only at considerably higher exposures (3.0 $mg \cdot kg^{-1} \cdot d^{-1}$ AP in pups and 30 $mg \cdot kg^{-1} \cdot d^{-1}$ in adult female rats).

Studies of perchlorate effects on thyroid function in mammalian wildlife models

In studies using deer mice, a wildlife model, Thuett et al. (2002a, 2002b) used AP exposures, the highest of which is estimated to be at least 4-fold higher than the highest exposure (30 $mg \cdot kg^{-1} \cdot d^{-1}$) used in the study of rats by York et al. (2001b). They exposed deer mice, throughout pregnancy and nursing, to 117.4 $ng \cdot kg^{-1} \cdot d^{-1}$, 117.4 $\mu g \cdot kg^{-1} \cdot d^{-1}$, 117.4 $mg \cdot kg^{-1} \cdot d^{-1}$ AP and sampled the pups at postnatal day 21. The pups had significantly decreased colloid area (i.e., colloid depletion) and decreased numbers of thyroid follicles per unit area at the lowest and highest (117.4 $ng \cdot L^{-1}$ and 117.4 $mg \cdot L^{-1}$) exposures compared to the controls. These data suggest that the thyroid glands in these groups were exposed to increased TSH release from HPT axis activation. In contrast, the intermediate treatment group (117.4 $\mu g \cdot L^{-1}$) did not exhibit significant changes in these variables (Thuett et al. 2002a). However, despite the histological evidence of HPT axis activation resulting from decreases in circulating THs in some treatment groups in this study, there was no evidence of decreased circulating THs. Surprisingly, 2 groups (117.4 $ng \cdot L^{-1}$ and 117.4 $\mu g \cdot L^{-1}$) had significantly increased plasma T_4 concentrations (plasma T_3 concentrations were not significantly altered). It seems likely that the elevated plasma T_4 concentrations observed in these groups reflects the release of stored hormone from the thyroid glands in response to increased TSH release (HPT axis activation). Such an "overshoot" in circulating hormones, as part of cyclic responses of HPT axis activation, is discussed in Chapter 3. A similar pattern of increased plasma T_4 concentrations, despite depleted thyroid gland hormone content, has been observed in some cases in avian species exposed to relatively low levels of AP but not when birds are exposed to very high concentrations of perchlorate in their drinking water (see Chapter 5).

The effects of perchlorate on thyroid function in prairie voles (*Microtus ochrogaster*), a herbivorous wildlife species, has been evaluated by measuring thyroid gland hormone content as well as plasma TH concentrations (Isanhart et al. 2005). Voles were exposed to 0, 1, or 10 $mg \cdot kg^{-1} \cdot d^{-1}$ AP in drinking water for 51 d, and another group of voles was exposed to 0.75 $mg \cdot kg^{-1} \cdot d^{-1}$ for 180 d. In the 51-d study, mean plasma T_4 concentrations decreased in a dose-dependent, but not significantly different, manner (Figure 4-3A), and plasma T_3 also was markedly decreased with increasing perchlorate exposure. This decrease in plasma T_3 is surprising because increased T_3/T_4 ratios in plasma (Delange and Ermans 1996) and the thyroid gland (Taurog 1996) are often observed with iodine deficiency. These alterations in the balance between T_3 and T_4 appear to reflect both increases in the ratio of T_3/T_4 produced by the thyroid when iodine is limiting, as well as increased deiodination within the thyroid gland when TSH secretion is increased (Rapoport and Spaulding 1996; Taurog 1996). Isanhart et al. (2005) found that although plasma T_4 in voles was

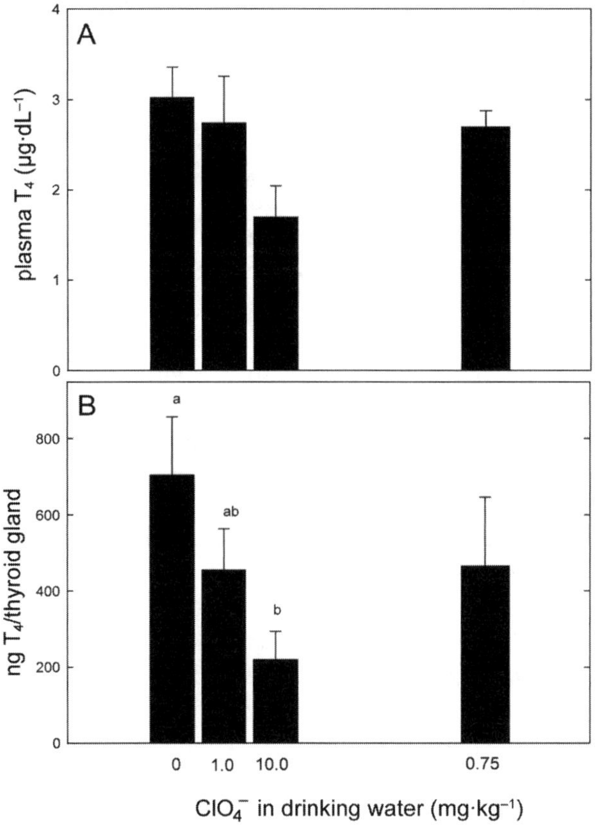

Figure 4-3 Plasma T_4 concentrations (A) and thyroid gland T_4 content (B) per treatment group of voles after 51 d of exposure to 0, 1.0, and 10.0 mg · kg^{-1} (n = 7/treatment) or 180 d to 0.75 mg · kg^{-1} (n = 5) in drinking water. Values are means ± standard error. Bars with different letters differ significantly ($p < 0.05$) (Reprinted with permission from Isanhart et al. 2005; copyright Society of Environmental Toxicology and Chemistry [SETAC]).

not significantly decreased by either of the treatments (1.0 and 10 mg·kg^{-1}·d^{-1}), thyroid gland T_4 content was decreased significantly in the group exposed to 10.0 mg·kg^{-1}·d^{-1} AP compared to controls (Figure 4-3B). Thus, measurement of thyroidal T_4 content revealed significant alterations in thyroid function of voles at perchlorate exposure levels where plasma T_4 was not significantly altered. Thus at this perchlorate exposure, the voles are depleting their thyroid gland stores of hormone to help maintain circulating T_4 concentrations (note that mean plasma T_4 tended to be decreased in this group although it was not significantly different from controls; Figure 4-3A). Although circulating TH concentrations are generally used as an indicator of perchlorate-induced alterations in thyroid function among laboratory and wildlife species (Thuett et al. 2002b; McNabb et al. 2004; York et al. 2004; Isanhart et al. 2005), thyroidal hormone content may better reflect the picture of sustained thyroid function. This method of assessment of thyroid function was developed in birds, and like thyroid histology, it appears to be a more sensitive measure of perchlorate effects than circulating TH concentrations (see Table 3-1; Chapter 5; McNabb 2005).

A second experiment was carried out with voles for 180 d at 0.75 mg perchlorate $\cdot$ kg^{-1} $\cdot$ d^{-1} (Isanhart et al. 2005). Plasma and glandular concentrations from this long-term exposure at 0.75 mg $\cdot$ kg^{-1} were comparable to T_4 concentrations from the 1 mg $\cdot$ kg^{-1} treatment group exposed for 51 d (Figure 4-3). The evaluation of thyroidal TH content and thyroid gland morphology provides a basis for the determination of the impact of perchlorate on thyroid gland function specifically, as well as the response of the HPT axis in "attempting" to maintain an organismal euthyroid state. These endpoints can be used to delineate dose–response relationships for perchlorate exposure and can be used to determine whether there are age-dependent differences in the effects of perchlorate in mammals.

As outlined in Chapter 3 and in the preceding section, TSH is the primary controller of the main aspects of thyroid gland function (iodide uptake, TH synthesis, and TH release). The thyroid gland also grows (hypertrophies) in response to increased TSH, leading to sustained changes in gland histology. Long-term exposure to perchlorate could result in sustained negative feedback, high TSH secretion, and the potential for tumor initiation and promotion in wildlife, as has been shown in laboratory species (see "Studies of perchlorate effects on thyroid function in laboratory mammals, " p 84).

Target Organ and System Effects of Perchlorate-induced Hypothyroidism in Mammals

Body and organ weight

In the study of Thuett et al. (2002b) in the section entitled "Studies of perchlorate effects on thyroid function in mammalian wildlife models," deer mice exposed to perchlorate during gestation, nursing, and post weaning showed perchlorate-associated body weight changes. Mean birth weight of the control deer mice in this study was similar to weights at parturition that had been established previously for the species (Layne 1968). Body weights in the midrange group (117.4 µg $\cdot$ L^{-1} AP in the drinking water) were consistently lower than in the control group and in other treatment groups after postnatal day 1 (PND 1), but this difference was significant only on PNDs 5 and 20. The body weights of the 117.4 mg $\cdot$ L^{-1} treatment group tended to be higher than those of controls.

As discussed in Chapter 3, THs are important for normal body and skeletal muscle development. Soukup and Jirmanova (2000) have demonstrated that THs influence the relative timing of muscle maturation and that certain tissue-specific threshold levels of THs are required for normal development (Kelly and Rubinstein 1986; Everts 1996). Male Wistar rats dosed with AP for 14 d at 10 $mg \cdot kg^{-1} \cdot d^{-1}$ exhibited decreased plasma TH concentrations but showed no differences in body or organ weights (Siglin et al. 2000). AP had no effect on body weights of fetuses, pups, or adult rats at exposures as high as 30.0 $mg \cdot kg^{-1} \cdot d^{-1}$ in maternal drinking water, during gestation and lactation, in 3 studies (York et al. 2001b, 2003, 2004). They found no general toxicity that could be attributed to exposure to perchlorate in the parent generation and no significant differences in average body weight with perchlorate exposure.

Liver weights at PND 21 of deer mice exposed to perchlorate during gestation and lactation via mothers drinking 117.4 $mg \cdot L^{-1}$ were significantly higher than those in the 117.4 $\mu g \cdot L^{-1}$ treatment group (Thuett et al. 2002b). However, when these data were analyzed with body weight as a covariate, there were no significant differences. No significant liver weight difference was observed between males and females.

Reproductive effects

Perchlorate effects on thyroid function might be expected, in turn, to affect reproductive function in mammals. A study by Buzzard et al. (2000) showed that TH receptors are expressed in developing rat testes, specifically in Sertoli cells which proliferate only during the fetal and early neonatal periods before they reach terminal differentiation. T_3 plays a crucial role in the development of Sertoli and Leydig cells, 2 primary cell types of the seminiferous tubule that are responsible for spermatogenesis and steroidogenesis in the testis (Cooke et al. 2004). Therefore, reductions of circulating THs resulting from perchlorate exposure could result in alteration of testicular size, number of Sertoli cells, and sperm production. In addition, experimental data indicate that T_3 acutely stimulates testosterone production by Leydig cells in vitro by stimulating steroidogenic acute regulatory protein (StAR) and StAR mRNA (Mendis-Handagama and Ariyaratne 2004). These thyroid-dependent phenomena could be affected by perchlorate and could potentially be responsible for the delayed time to conception observed in some of the perchlorate-treated animals (Jannini et al. 1995).

In studies of deer mice and rats (Thuett et al. 2002b; York et al. 2004, respectively), exposure to perchlorate during gestation and lactation did not affect pups survival or their later reproductive performance (litter size or viability index; deer mice, Thuett et al. 2002b; rats, York et al. 2004). Testicular morphological changes were observed in perchlorate-treated rodents; specifically, there was an increase in seminiferous tubule diameter of the testes of deer mice (Thuett et al. 2002a).

Nervous system effects

Alterations in morphology of the brain have been suggested in a rodent model following exposure to perchlorate during gestation and lactation (York et al. 2004). Perchlorate tended to increase the size of the corpus callosum in female pups following exposure of the dams to 10 mg AP salt $\cdot$ kg^{-1} $\cdot$ d^{-1} during gestation and lactation. Although the mean corpus callosum weight was 23.3% greater in perchlorate-treated animals than in controls, this difference was not statistically significant. Although these trends are suggestive of effects on the brain resulting from perchlorate treatment, these studies need to be repeated. Previous studies of perchlorate have shown that hypothyroidism induced by antithyroid agents or by severe iodine deficiency can alter the size of the brain in rats (York et al. 2001b). There are critical periods for TH action on the developing brain, and reduced levels of THs at that time in development can result in deficits of specific neurological functions including visual processing, motor skills, and responses (Zoeller and Rovet 2004). While the specific mechanisms for these deficits have not been clearly determined, it is possible that under ecotoxicological conditions, such neurological deficits could lead to poor survival rates.

Metabolism and thermoregulation

Isanhart et al. (2005) recently evaluated the effects of perchlorate (0, 1 and 10 mg$\cdot$kg$^{-1}\cdot$d^{-1}) on resting metabolic rates, facultative thermogenesis (peak metabolic rate), and thyroid function in prairie voles (see "Studies of perchlorate effects on thyroid function in mammalian wildlife models," p 85). They found that exposure to perchlorate for 51d or 180 d had no effect on resting metabolic rate or facultative thermogenesis in the prairie vole. Although perchlorate exposure (10 mg$\cdot$kg$^{-1}\cdot$d^{-1}) significantly decreased thyroid gland hormone content, plasma TH concentrations were not significantly decreased. Thus, the exposure concentrations and exposure times used in this study did not cause significant organismal hypothyroidism; therefore, the lack of effects on metabolism may not be surprising. It is not expected that decreased metabolism would be apparent unless circulating THs were also significantly decreased. Because the mammalian thyroid gland can store large amounts of hormone (Guyton and Hall 2001), physiological effects of reduced TH production may not be observed for weeks or months. Thus, a longer duration of exposure may be necessary to induce any possible perchlorate-related reductions in resting or peak metabolic rate.

Ecotoxicology of Perchlorate Exposure in Mammals

The most extreme examples of perchlorate contamination are associated with ground and surface bodies of water located near military bases and ammunitions manufacturing sites where perchlorate is handled. However, these sources of perchlorate may lead to more widespread contamination if they feed into large bodies of water or underground aquifers. Perchlorate contamination at low levels is found in some arid areas and in association with perchlorate-contaminated fertilizer use. (The environmental distribution of perchlorate contamination has been discussed in Chapter 1.) Perchlorate exposure of mammals is most likely to occur via the intake of contaminated drinking water or food, or through the intake of other extraneous material (such as soil) with the food. Thus, the ecotoxicological risk to mammalian wildlife will depend not only on the extent of environmental contamination with perchlorate but also on the nature of water and food intake in the context of the life history characteristics of individual species.

Although most mammalian wildlife species are terrestrial, those living in aquatic ecosystems at the terrestrial–aquatic ecosystem interface in regions of high perchlorate contamination are likely to be at the greatest risk for perchlorate exposure and subsequent adverse physiological responses. Those mammals that have high drinking rates (high water turnover), for example, species with small body size (water turnover is proportional to size), those that have high nitrogen intake (which requires water use in excretion), and those that use significant amounts of water for evaporative cooling, will have proportionally higher perchlorate intake in a contaminated environment than will mammals with lower drinking rates (for a general discussion of drinking and water turnover in mammals, see Vaughn 1986). Herbivorous and granivorous terrestrial mammals also may take in perchlorate by eating plants that accumulate perchlorate, or they may ingest perchlorate-contaminated soil with their food. Carnivorous mammals may eat vertebrates or invertebrates that contain perchlorate (for examples, see Smith et al. 2001, 2004). Developing mammals are exposed to perchlorate by placental transfer relative to the perchlorate exposure of the mother, and in the postnatal period, mammalian young are exposed via the mother's milk (Kirk et al. 2005).

Some studies have attempted to assess perchlorate exposure and consequences in wild and domestic animals living in areas of environmental perchlorate contamination. Both Smith et al. (2005) and Cheng et al. (2004) hypothesized that plasma TH concentrations of raccoons and cows, respectively, would be altered in response to environmental perchlorate exposure. Raccoons captured at the LHAAP, in areas with potential food items containing up to 5500 $\mu g \cdot g^{-1}$ perchlorate ion, did not have detectable blood perchlorate (detection limit of 30 $\mu g \cdot L^{-1}$), so their perchlorate exposure could not be confirmed. There were no significant differences in plasma T_3 or TSH concentrations between raccoons captured near the Harrison Bayou area (the most perchlorate-contaminated area in LHAAP) and those from other areas of LHAAP. Plasma TH concentrations also were not inversely related to plasma TSH concentrations as would have been expected if perchlorate exposure were decreasing thyroid function. The authors suggest that one aspect of the plasma hormone data was indicative of a perchlorate effect; plasma T_3 concentrations from raccoons were significantly different between 2 different years (2000, mean = 27.3 $ng \cdot dl^{-1} \pm 2.6$ SE; and 2001, mean = 44.1 $ng \cdot dl^{-1} \pm 2.9$ SE). The authors speculate that the increase in plasma T_3 from 2000 to 2001 could reflect decreasing perchlorate exposure resulting from the installation of a fluidized bed reactor bioremediation system for removing perchlorate prior to water discharge into Harrison Bayou that was installed in late 2000 between the times of these 2 samplings. In a separate study, Cheng et al. (2004) compared 2 heifer calves that drank ~25 ng perchlorate ion $\cdot$ mL^{-1} with 2 heifers that drank uncontaminated water. Plasma samples at 2-week intervals over the 14 weeks of the study showed detectable plasma perchlorate concentrations in 1 perchlorate-exposed heifer on 2 occasions but no other detectable perchlorate in the plasma samples (detection limit for plasma perchlorate was 13.7 $ng \cdot mL^{-1}$). Plasma T_3 and T_4 concentrations were not altered by perchlorate exposure, although circulating TH concentrations tended to be lower in the cows that drank perchlorate. Overall, this study gave no indication of alterations in organismal thyroid status after cattle had been exposed to 25 $\mu g \cdot L^{-1}$ perchlorate in drinking water for 14 weeks. These exposures are much lower than those that caused hypothyroidism in rats in laboratory studies.

Physiologically Based Pharmacokinetic Modeling

Physiologically based pharmacokinetic modeling (PBPK) has been used to predict the effects of perchlorate on laboratory rats (Clewell et al. 2003a, 2003b). Merrill et al. (2003) developed a PBPK model to predict the effects of perchlorate on iodide uptake in the male rat, based on a preliminary model by Fisher et al. (2000). The model describes simultaneous kinetics for both radioiodide (I^-) and perchlorate (ClO_4^-), together with their interaction at the sodium–iodide symporter (NIS). The compartments for the NIS-containing tissues, stomach, thyroid, and skin, were divided into subcompartments. The stomach compartment was divided into gut contents, gut wall, and capillary bed. For iodide, the thyroid compartment was divided into the stroma, the follicle, the colloid, as well as a separate compartment representing iodide binding to TH. For perchlorate, only the stroma, follicle, and colloid subcompartments comprised the thyroid compartment. The skin compartment was divided into skin tissue and capillary bed. The model predicts serum and thyroid ClO_4^- concentrations from 14-d drinking water exposures ranging from 0.01 to 3.0 $mg \cdot kg^{-1}$ and also successfully predicts perchlorate's interaction with iodide. The model was the basis for a second model used to predict fetal perchlorate dose and inhibition of iodide kinetics during gestation (Clewell et al. 2003b). The Clewell model differs from the Merrill model primarily in the addition of a fetus PBPK model similar to the Merrill model and the addition of mammary gland and placenta compartments in the adult pregnant dam model.

A PBPK model to predict the effects of environmental exposures of ClO_4^- on mammalian wildlife is currently not available. Development of a PBPK model with TH submodels (DiStefano and Fisher 1976; Saratchandran et al. 1976; DiStefano and Mori 1977) to predict the uptake and distribution of perchlorate in body tissues and the effects of perchlorate on the TH system would be beneficial for mammalian ecotoxicological simulations and predictions. Considering environmental distribution of perchlorate as discussed in Chapter 1, exposure scenarios as discussed in "Perchlorate Effects on Thyroid Function" (p 84), and life history characteristics of individual species, this model would provide exotoxicological data that cannot be obtained from laboratory animal models or exposure conditions.

Summary

It is clear that the primary target of perchlorate is the thyroid gland. Most knowledge of perchlorate effects in mammals is limited to data on the effects on THs and thyroid gland histology. The thyroid gland–TH content, like thyroid histology, is an especially sensitive indicator of perchlorate effects, in contrast to circulating THs and thyroid gland weights. The sensitivity of plasma TSH changes as an indicator of perchlorate's effect on the thyroid is variable among different studies. Almost all work on the effects of perchlorate on mammals has been on laboratory rats, although a few studies have used wildlife models in the laboratory. Recently, 2 studies of mammals (racoons and cattle) exposed to perchlorate in the wild have shown minimal effects of perchlorate on thyroid function.

Ecotoxicological risks to mammalian wildlife will depend not only on the extent of environmental contamination with perchlorate but also on the nature of water and food intake in the context of the life history characteristics of individual species. Mammalian species living near aquatic ecosystems and terrestrial–aquatic interfaces in regions of high perchlorate contamination are likely to be at the greatest risk for perchlorate exposure and subsequent adverse physiological responses. Herbivorous and granivorous terrestrial mammals can be exposed to perchlorate by consumption of contaminated plants, food, or soil. Carnivorous mammals may consume vertebrates or invertebrates contaminated with perchlorate. Prenatal and postnatal developing mammals are exposed to perchlorate by placental transfer and during lactation.

Research Needs: Perchlorate Studies in Mammals

- Environmentally relevant dose–response evaluations of perchlorate effects are needed to establish the association with thyroid disruption effects.
- Data on maternal exposure and milk transfer to neonatal and developing offspring are needed for risk characterization and assessment.
- Effects of exposure to increasing concentration of perchlorate in milk and decreasing concentration of iodide represent additional areas of research that need to be addressed.
- Definitive, hypothesis-driven evaluation of the metabolic and thermoregulatory impacts of perchlorate in various species should be initiated.
- Studies of wildlife, both in the laboratory and in nature, are needed because most current information is limited to laboratory models.
- The effects of perchlorate on cellular and molecular mechanisms of action need to be addressed, for example, the effects of perchlorate on transport proteins such as NIS and pendrin.
- A mammalian wildlife PBPK model for ecotoxicological simulations and predictions should be developed.

References

Beyer WN, Connor EE, Gerould S. 1994. Estimates of soil ingestion by wildlife. J Wildl Manag 582:375–382.

Buzzard JJ, Morrison JR, O'Bryan MK, Song Q, Wreford NG. 2000. Developmental expression of thyroid hormone receptors in the rat testis. Biol Reprod 62:664–669.

Capen CC. 1994. Mechanisms of chemical injury of thyroid gland. Prog Clin Biol Res 387:173–91.

Cheng Q, Perlmutter L, Smith PN, McMurry ST, Jackson WA, Anderson TA. 2004. A study on perchlorate exposure and absorption in beef cattle. J Agric Food Chem 52:3456–61.

Chow SY, Chang LR, Yen MS. 1969. A comparison between the uptakes of radioactive perchlorate and iodide by rat and guinea-pig thyroid glands. J Endocrinol 45:1–8.

Chow SY, Woodbury DM. 1970. Kinetics of distribution of radioactive perchlorate in rat and guinea-pig thyroid glands. J Endocrinol 47:207–218.

Clewell RA, Merrill EA, Yu KO, Mahle DA, Sterner TR, Fisher JW, Gearhart JM. 2003a. Predicting neonatal perchlorate dose and inhibition of iodide uptake in the rat during lactation using physiologically-based pharmacokinetic modeling. Toxicol Sci 74:416–436.

Clewell RA, Merrill EA, Yu KO, Mahle DA, Sterner TR, Mattie DR, Robinson PJ, Fisher JW, Gearhart JM. 2003b. Predicting fetal perchlorate dose and inhibition of iodide kinetics during gestation: a physiologically-based pharmacokinetic analysis of perchlorate and iodide kinetics in the rat. Toxicol Sci 73:235–255.

Cooke PS, Holsberger DR, Witorsch RJ, Sylvester PW, Meredith JM, Treinen KA, Chapin RE. 2004. Thyroid hormone, glucocorticoids, and prolactin at the nexus of physiology, reproduction, and toxicology. Toxicol Appl Pharmacol 194:309–35.

Delange FM, Ermans A-M. 1996. Iodine deficiency. In: Braverman LE, Utiger RD, editors. Werner and Ingbar's the thyroid, 7th ed. Philadelphia (PA): Lippincott-Raven. p 296–316.

DiStefano JJ III, Fisher DA. 1976. Peripheral distribution and metabolism of the thyroid hormones: a primarily quantitative assessment. Pharmacol Ther 2:539–570.

DiStefano JJ III, Mori F. 1977. Parameter identifiability and experiment design: thyroid hormone metabolism parameters. Am J Physiol 233:R134–144.

Everts ME. 1996. Effects of thyroid hormones on contractility and cation transport in skeletal muscle. Acta Physiol Scand 156:325–333.

Fisher J, Todd P, Mattie D, Godfrey D, Narayanan L, Yu K. 2000. Preliminary development of a physiological model for perchlorate in the adult male rat: a framework for further studies. Drug Chem Toxicol 23:243–258.

Guyton AC, Hall JE. 2001. The thyroid metabolic hormones. In: Textbook of medical physiology, 10th ed. Philadelphia (PA): WB Saunders. p 859.

Isanhart JP, McNabb FM, Smith PN. 2005. Effects of perchlorate exposure on resting metabolism, peak metabolism, and thyroid function in the prairie vole (*Microtus ochrogaster*). Environ Toxicol Chem 24:678–684.

Jannini EA, Ulisse S, D'Armiento M. 1995. Thyroid hormone and male gonadal function. Endocr Rev 16:443–459.

Kelly AM, Rubinstein NA. 1986. Development of neuromuscular specialization. Med Sci Sports Exerc 18:292–298.

Kessler FJ, Kruskemper HL. 1966. Experimental thyroid tumors caused by many years of potassium perchlorate administration. Klin Wochenschr 44:1154–1156.

Kirk AB, Martinelango PK, Tian K, Dutta A, Smith EE, Dasgupta PK. 2005. Perchlorate and iodide in dairy and breast milk. Environ Sci Technol 39:2011–2017.

Kirk AB, Smith EE, Tian K, Anderson TA, Dasgupta PK. 2003. Perchlorate in milk. Environ Sci Technol 37:4979–4981.

Layne JN. Ontogeny. 1968. In: King JA, editor. Biology of peromyscus (Rodentia). Stillwater (OK): American Soc of Mammologists. p 148–248.

Mendis-Handagama SM, Ariyaratne HB. 2004. Effects of thyroid hormones on Leydig cells in the postnatal testis. Histol Histopathol 19:985–997.

McNabb FMA. 2005. Biomarkers for the assessment of avian thyroid disruption by chemical contaminants. Poult Avian Biol Rev 16:3–10

McNabb FMA, Jang DA, Larsen CT. 2004. Does thyroid function in developing birds adapt to sustained ammonium perchlorate exposure? Toxicol Sci 82:106–113.

Merrill EA, Clewell RA, Gearhart JM, Robinson PJ, Sterner TR, Yu KO, Mattie DR, Fisher JW. 2003. PBPK predictions of perchlorate distribution and its effect on thyroid uptake of radioiodide in the male rat. Toxicol Sci 73:256–269.

Parsons, Inc. 2001. Scientific and technical report for perchlorate biotransport investigation: A study of perchlorate occurrence in selected ecosystems. Interim final report. Brooks Air Force Base (TX): Air Force Institute for Environment, Safety and Occupational Health Risk Analysis.

Pisarev MA, Kleiman de Pisarev DL. 1980. Biochemistry of thyroid regulation under normal and abnormal conditions. J Endocrinol Invest 3:317–329.

Pontius FW, Damian P, Eaton AD. 2000. Regulating perchlorate in drinking water. In: Urbansky ET, editor. Perchlorate in the environment. New York (NY): Kluwer Academic/Plenum. p 31–36.

Rapoport B, Spaulding SW. 1996. Mechanism of action of thyrotropin and other thyroid growth factors. In Braverman LE, Utiger RD, eds, Werner and Ingbar's the thyroid, 7th ed. Philadelphia (PA): Lippincott-Raven. p 207–219.

Saratchandran P, Carson ER, Reeve J. 1976. An improved mathematical model of human thyroid hormone regulation. Clin Endocrinol (Oxf) 5:473–483.

Siglin JC, Mattie DR, Dodd DE, Hildebrandt PK, Baker WH. 1998. A 90-day drinking water toxicity study in rats of the environmental contaminant ammonium perchlorate. Toxicol Sci 57:61–74.

Smith PN, Theodorakis CW, Anderson TA, Kendall RJ. 2001. Preliminary assessment of perchlorate in ecological receptors at the Longhorn Army Ammunition Plant (LHAAP), Karnack, Texas. Ecotoxicology 10:305–313.

Smith PN, Utley SJ, Cox SB, Anderson TA, McMurry ST. 2005. Monitoring perchlorate exposure and thyroid hormone status among raccoons inhabiting a perchlorate-contaminated site. Environ Monit Assess 102:337–347.

Smith PN, Yu L, McMurry ST, Anderson TA. 2004. Perchlorate in water, soil, vegetation, and rodents collected from the Las Vegas Wash, Nevada, USA. Environ Pollut 132:121–127.

Soukup T, Jirmanova I. 2000. Regulation of myosin expression in developing and regenerating extrafusal and intrafusal muscle fibers with special emphasis on the role of thyroid hormones. Physiol Res 49:617–633.

Taurog A. 1996. Hormone synthesis: Thyroid iodine metabolism. In Braverman LE, Utiger RD, eds, Werner and Ingbar's The Thyroid. 7th edition. Lippincott-Raven, Philadelphia 47-81.

Thuett KA, Roots EH, Mitchell LP, Gentles BA, Anderson T, Kendall RJ, Smith EE. 2002a. Effects of in utero and lactational ammonium perchlorate exposure on thyroid gland histology and thyroid and sex hormones in developing deer mice (*Peromyscus maniculatus*) through postnatal day 21. J Toxicol Environ Health A 65:2119–2130.

Thuett KA, Roots EH, Mitchell LP, Gentles BA, Anderson TA, Smith EE. 2002b. In utero and lactational exposure to ammonium perchlorate in drinking water: effects on developing deer mice at postnatal day 21. J Toxicol Environ Health A 65:1061–1076.

Vaughn TA. 1986. Water regulation. In: Mammalogy, 3rd ed. Philadelphia (PA): Saunders College Publ. 477–493.

Wolff J. 1998. Perchlorate and the thyroid gland. Pharmacol Rev 50:89–105.

York RG, Barnett Jr J, Brown WR, Garman RH, Mattie DR, Dodd D. 2004. A rat neurodevelopmental evaluation of offspring, including evaluation of adult and neonatal thyroid, from mothers treated with ammonium perchlorate in drinking water. Int J Toxicol 23:191–214.

York RG, Brown WR, Girard MF, Dollarhide JS. 2001a. Oral (drinking water) developmental toxicity study of ammonium perchlorate in New Zealand White rabbits. Int J Toxicol 20:199–205.

York RG, Brown WR, Girard MF, Dollarhide JS. 2001b. Two-generation reproduction study of ammonium perchlorate in drinking water in rats evaluates thyroid toxicity. Int J Toxicol 20:183–197.

York RG, Funk KA, Girard MF, Mattie D, Strawson JE. 2003. Oral (drinking water) developmental toxicity study of ammonium perchlorate in Sprague-Dawley rats. Int J Toxicol 22:453–464.

Zoeller RT, Rovet J. 2004. Timing of thyroid hormone action in the developing brain: clinical observations and experimental findings. J Neuroendocrinol 16:809–818.

five

Perchlorate Effects on Birds

FM Anne McNabb, Michael J Hooper, Ernest E Smith, Scott McMurry, B Angella Gentles

Introduction to Studies of Perchlorate Effects on Birds

Historical studies

There are few studies in the literature on the effects of perchlorate on avian thyroid function. This seems surprising because perchlorate has often been used as an experimental goitrogen in mammals and was used as a thyroid inhibitor in cases of human hyperthyroidism. These uses of perchlorate in clinical medicine and laboratory mammals have provided a considerable amount of basic information about the mechanism of perchlorate action on mammalian thyroid function (Chapters 3 and 4). Considering that perchlorate is a thyroid inhibitor with a single primary mode of action (inhibition of thyroidal I^- transport), it is surprising that this anion has not been used routinely in basic studies of avian thyroid function.

Several studies have evaluated perchlorate uptake and its effects on I^- dynamics in birds. Labeled perchlorate uptake, into the thyroid gland and other I^--transporting tissues of adult chickens, was found to be very similar to the pattern of radiolabelled I^- uptake in those same tissues (Peña et al. 1975, 1976). Peña et al. (1966) found that the uptake of I^- into extrathyroidal avian tissues such as ovaries, which concentrate I^-, also was inhibited by perchlorate. In their study, potassium perchlorate exposure of hens (at 200 mg per bird per day) decreased radiolabelled I^- deposition in eggs to 9% of that in eggs from control hens. Radiolabelled perchlorate ($K^{36}ClO_4$)

Perchlorate Ecotoxicology. Ronald J Kendall and Philip N Smith, editors.

ISBN-10: 1-880611-90-2 ISBN-13: 978-1-880611-90-6

exposure of hens also resulted in perchlorate deposition in eggs (Peña et al. 1976). Perchlorate treatment in ovo (5 mg/egg potassium perchlorate) decreased the thyroid gland content of I^- of fowl hatchlings from treated eggs, although thyroid weights and hatching time did not differ significantly from controls. This lack of alteration in hatching time suggested that circulating thyroid hormones in the embryos were not altered by the perchlorate treatment (Brown et al. 1991). Although these studies indicated that some basic components of perchlorate effects in birds are similar to those in mammals, they provided little specific information about the effects of perchlorate on avian thyroid function.

Implications of perchlorate exposure

Life stages

The young of oviparous animals develop in isolated eggs in which the mother has deposited a variety of materials that influence embryonic development (review, McNabb et al. 1999). In the context of perchlorate effects on avian embryos, the key components that need to be considered are the deposition of I^-, maternal thyroid hormones (THs) and perchlorate. Some birds are very resistant to developing hypothyroidism. For example, Japanese quail (*Coturnix coturnix japonica*) hens remained euthyroid (based on their circulating TH concentrations) after 56 d on a purified diet with no detectable I^- in the components, although by this time their eggs were I^- deficient (McNabb et al. 1985). If the hen is mildly hypothyroid as a result of perchlorate exposure, her eggs are likely to have reduced TH content because the deposition of maternal THs into eggs is in proportion to the mother's thyroid status (Wilson and McNabb 1997). However, when female birds are severely hypothyroid, they cease laying, so it is unlikely that eggs with very low maternal TH content will be produced (Wilson and McNabb 1997). It is not understood how important maternal THs are in embryonic development in the period prior to the onset of appreciable functional capacity of the embryonic thyroid gland. However, there is evidence for altricial rats that maternal THs are critical for early development prior to the beginning of thyroid function (Morreale de Escobar et al. 1985). Birds exposed to perchlorate, in addition to having low TH content in their eggs, also will have decreased egg iodine (see "Historical studies," p 99 and "Perchlorate deposition in eggs," p 103) and increased egg perchlorate concentrations (Peña et al. 1976). Thus, deficient iodine stores and perchlorate inhibition of thyroidal I^- uptake by the embryonic thyroid gland both will decrease TH synthesis. If deficient TH synthesis results in embryonic hypothyroidism, this may affect a number of developmental events that are under the control of THs as well as having detrimental effects on hatching (review, McNabb 2000).

Modes of development

It seems likely that perchlorate effects in embryonic and early posthatch life will be different in precocial and altricial birds. In precocial birds, which are hatched at a relatively advanced stage of physiological development, the thyroid gland is organized into follicles, is producing significant amounts of hormone and comes under hypothalamic–pituitary–thyroid (HPT) axis control by about the middle of embryonic life (Thommes 1987). In embryos, circulating concentrations of THs are low for about the first half of incubation, they increase by 3 to 4 fold by the beginning of the perihatch period, then peak dramatically during the perihatch period, and decrease after hatch (Figure 5-1; reviews, McNabb and Olson 1996; McNabb et al. 1998). THs are required for the development of a number of systems that are relatively mature at the time of hatch in precocial species (e.g., central nervous system, muscle, musculoskeletal system, gut; review, McNabb and King 1993). Thus, in precocial species, the increasing thyroid function during late embryonic life appears to be critical to the developmental maturation of these systems, and the peak of circulating THs during the perihatch period appears to be important in the initiation of thermoregulatory responses at hatch (review, McNabb

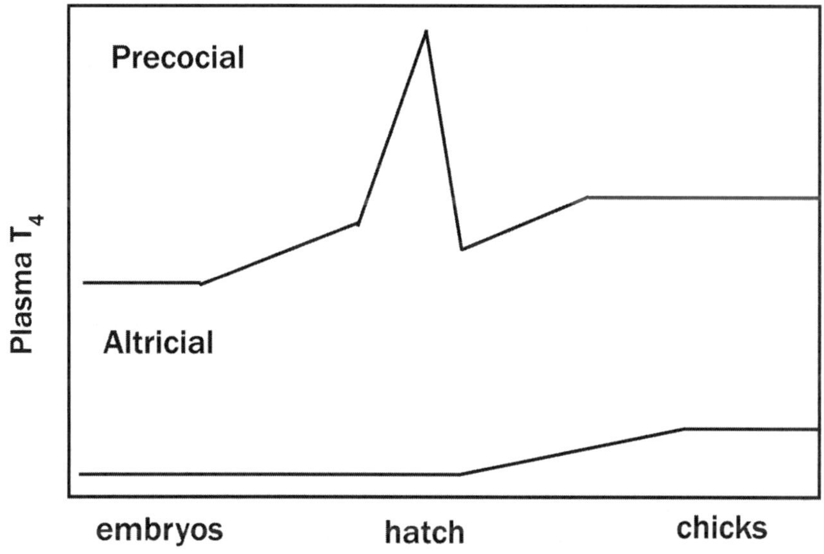

Figure 5-1 Circulating thyroxine concentrations during development in precocial and altricial birds

and Olson 1996). Embryos in eggs laid by perchlorate-exposed birds may experience decreased maternal TH stores, decreased iodine availability, and additional inhibition of thyroidal I^- uptake due to egg perchlorate. If these factors result in embryonic hypothyroidism, it is likely to alter embryonic development of many body systems and delay or interfere with the hatching process.

In contrast, altricial birds are hatched at a less mature state and remain essentially helpless and confined to the nest for some time. Altricial nestlings are poikilothermic for most of their nestling life, and dependent on parental care until fledging. There is little thyroid functional development until after hatch in altricial species; plasma THs are very low during embryonic life and the perihatch period, and HPT axis control of thyroid function is not established until after hatch (McNichols and McNabb 1988). Thyroid function gradually increases during nestling life, preceding the development of thermoregulatory capability just before fledging (reviews, McNabb and Olson 1996; McNabb et al. 1998).

It appears that altricial embryos should be less affected by ammonium perchlorate in ovo than precocials because there is very little thyroid development in altricials during embryonic life and their embryonic development appears to be largely TH independent. Likewise, altricial species might be expected to be more affected by perchlorate exposure during posthatch development when most thyroid function is developing. Chicks hatched from eggs of perchlorate-exposed hens may experience relatively low THs, low I^-, and high perchlorate in ovo. After hatch, both altricial nestlings and precocial chicks are likely to gain I^- from the food, so the ratio of perchlorate to I^- may be altered. Sufficient or high I^- availability is known to lessen the effects of perchlorate at least in mammals (Taurog 1996), so there may be some differences in the relative perchlorate effects before and after hatch due to differences in I^- availability.

Routes of exposure

The nature of the habitat in which birds live, the extent of its perchlorate contamination, and the bioavailability of perchlorate will be important factors in avian exposure to environmental perchlorate. In addition, it is necessary to understand the route of ingestion, the nature of the diet and foraging patterns, and the magnitude of water turnover of the species under consideration. The relative importance of perchlorate input from water versus food will differ between species. These topics will be discussed further in an ecological context in "Ecotoxicology of Perchlorate Exposure in Birds," p 114.

General Picture of Perchlorate Effects on Birds

Recent work indicates that, in general, perchlorate effects on thyroid function in birds are similar to those in mammals (Chapters 3 and 4). This is not surprising, considering the extent to which the key aspects of thyroid function are similar throughout the vertebrate classes and because perchlorate appears to have a single primary mechanism of action, competitive inhibition of thyroidal I^- uptake. However, only a few recent investigations of perchlorate effects are sufficiently detailed to allow assessment of the similarities and differences between perchlorate effects among the different vertebrate classes. Those investigations relevant to birds will be discussed below.

Laboratory Studies of Effects of Perchlorate in Birds

Perchlorate deposition in eggs

A recent study of bobwhite quail (*Colinus virginianus*) hens exposed to ammonium perchlorate in drinking water has demonstrated perchlorate deposition into eggs (Gentles et al. 2005). In this study, hens were exposed to ammonium perchlorate in drinking water at 1.18, 11.75 and 117.5 $mg \cdot L^{-1}$. Egg perchlorate ion concentrations increased significantly with increasing ammonium perchlorate exposure (Figure 5-2). The eggs from control hens had 0.037 ± 0.024 (mean ± SE) $\mu g \cdot g^{-1}$ perchlorate ion;

- at 1.18 $mg \cdot L^{-1}$ ammonium perchlorate in the water, egg perchlorate was 0.599 ± 0.063 $\mu g \cdot g^{-1}$ perchlorate ion;
- at 11.75 $mg \cdot L^{-1}$ ammonium perchlorate, the mean was 3.784 ± 0.423 $\mu g \cdot g^{-1}$ perchlorate ion, and
- at 117.5 $mg \cdot L^{-1}$ ammonium perchlorate the mean was 7.158 ± 0.966 $\mu g \cdot g^{-1}$ perchlorate ion.

The numbers of eggs laid, the egg dimensions, and the egg weight were not affected by up to 30 days of exposure in this study. There were fluctuations in egg production in this study (i.e., a significant date effect) but no indication that these fluctuations were related to perchlorate exposure. The data on egg perchlorate deposition (Gentles et al. 2005) augment those of Peña et al. (1976), which showed that radiolabelled perchlorate administered to laying hens was deposited in their ova. The effects of perchlorate on thyroid histology of hens in this study will be discussed in "Metabolism and thermoregulation," p 114.

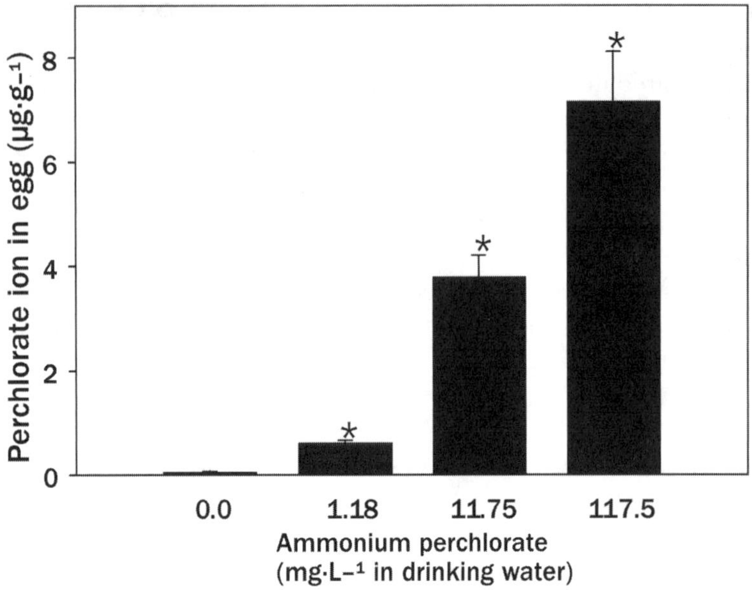

Figure 5-2 Perchlorate ion concentration in eggs laid by bobwhite quail hens exposed to ammonium perchlorate in drinking water. Asterisks denote significant differences from control. Values are means ± SE. (Reprinted with permission from Gentles et al. 2005; copyright Society of Environmental Toxicology and Chemistry [SETAC])

Effects of egg perchlorate on embryonic thyroid function

The only studies of perchlorate effects on embryonic development are in bobwhite quail and mallard duck embryos exposed to ammonium perchlorate in ovo. Both of these U.S. Environmental Protection Agency (USEPA)-approved model species are precocial with considerable thyroid development during embryonic life (see "Modes of development," p 101), so their thyroid development and function would be likely to be affected by in ovo perchlorate exposure. In these studies, ammonium perchlorate was injected into the air cell on day 4 of incubation, and embryos were sampled during the latter part of incubation but prior to the beginning of the perihatch peak of circulating THs (on day 21 of the 23 to 24 d incubation period in quail or on day 24 of the 28 d incubation period in mallards). At this time in embryonic development, thyroid gland function is relatively mature, and perchlorate exposure should decrease TH production and release and cause a feedback response by the HPT axis. In these 21 d quail embryos, egg ammonium perchlorate concentrations of 0.025 to 300 $\mu g \cdot g^{-1}$ did not result in dose-related effects on plasma THs or thyroid weight (HPT axis activation) as was predicted. However, thyroid function was altered because thyroidal T_4 content of embryos was significantly decreased at ≥0.5 $\mu g \cdot g^{-1}$ egg ammonium perchlorate (~0.43 $\mu g \cdot g^{-1}$ egg perchlorate ion). Some ammonium perchlorate-treated eggs were allowed to hatch, and the effects observed for embryos were sustained in hatchlings less than 1 day of age (McNabb et al. 2002, McNabb 2004). The studies described above indicate that in ovo perchlorate concentrations that alter thyroidal TH content in bobwhite embryos are less than those measured by Gentles et al. (2005) in eggs from hens drinking 1.18 $mg \cdot L^{-1}$ ammonium perchlorate or higher (see "Perchlorate deposition in eggs," p 103). However, these studies address only the effects of perchlorate on embryos. Eggs laid by perchlorate-treated hens also may be deficient in I^- and maternal THs, so the impact on embryos should be greater than with perchlorate exposure alone.

Studies of mallard embryos, exposed in ovo to 0.025 $\mu g \cdot g^{-1}$ to 240 $mg \cdot g^{-1}$ ammonium perchlorate, indicate plasma THs were not significantly decreased and thyroid weights were not increased by perchlorate exposure. However, thyroidal T_4 content was decreased with increasing perchlorate exposure; the values were significantly less than those of controls at ≥0.333 $mg \cdot g^{-1}$ egg ammonium perchlorate. Thus, it appears that thyroid function in mallard embryos is much less sensitive to ammonium perchlorate than it is in bobwhite quail embryos; the ammonium perchlorate concentration in eggs that caused thyroidal T_4 depletion was much higher for mallard than for quail embryos. Both quail and mallard embryos, by about 3/4 of the way through incubation (the time of sampling), can maintain circulating TH concentrations at euthyroid levels during exposure to the range of ammonium perchlorate concentrations used in these studies. The general picture of responses, that is, that thyroidal T_4 content is depleted at ammonium perchlorate exposures below those that affect plasma THs or thyroid gland weights, is similar in quail and mallard embryos (McNabb et al. 2002; McNabb 2004).

Effects of perchlorate on chick and adult thyroid function

Assessments of thyroid disruption

Recently, the effects of ammonium perchlorate in drinking water on thyroid function and its HPT axis control have been studied in chicks and adult bobwhite quail and in mallard ducklings. As mentioned above, the chicks or ducklings of these species are precocial, and thus their thyroid function and general state of maturation are relatively advanced at the time of hatch. Thyroid function has been assessed in 3 ways in these studies:

1) plasma THs as an index of organismal thyroid status,
2) thyroid gland weights as an index of HPT axis activation, and
3) thyroid gland–TH content (thyroidal TH) as an index of hormone stores.

Comparison of these thyroid variables indicates that thyroidal T_4 content is by far the most sensitive indicator of altered thyroid function; plasma hormones and thyroid gland weights are less sensitive and approximately equal as indicators of altered thyroid function (McNabb, Larsen, Pooler 2004; and see Table 3-1, Chapter 3). Measurement of both thyroidal T_4 and T_3 indicate that T_4 comprises >95% of gland hormone content for a wide range of ammonium perchlorate exposures, so T_3 measurements add little additional information. Studies of perchlorate exposure in embryos, chicks, and adults of bobwhite quail; in mallard embryos and ducklings; and in herring gulls exposed to polyhalogenated aromatic hydrocarbons, all support the idea that thyroidal T_4 is the most sensitive indicator of altered thyroid function resulting from contaminant chemicals (McNabb 2005).

In studies of quail chicks exposed to perchlorate, plasma THs were the most variable index measured. At relatively low perchlorate concentrations in the drinking water, plasma THs were sometimes increased and sometimes decreased compared to controls. This pattern appears to be related to cyclic responses that are characteristic of iodine-deficient conditions. In brief, a decrease in TH synthesis and release leads to decreases in circulating THs that normally exert a negative feedback effect on the HPT axis. The consequent increase in thyroid-stimulating hormone (TSH) secretion stimulates the release of stored hormone from the thyroid gland and restores euthyroid concentrations of THs in the circulation (or sometimes even leads to an "overshoot" effect), thereby relieving the feedback effect on the HPT axis. If the cause of the decreased thyroid function persists, this pattern may repeat, resulting in cyclic fluctuations in circulating hormone concentrations. Such cyclic responses may be a cause of the high variability in plasma TH concentrations between individuals (McNabb, Jang, Larsen 2004; McNabb, Larsen, Pooler 2004). Cyclic patterns of circulating THs are known to occur in humans in response to iodine deficiency (Delange and Ermans 1996), a situation comparable to that resulting from perchlorate exposure.

Bobwhite chicks

Relatively short-range exposures of chicks to ammonium perchlorate (2 weeks, starting at days 3 to 4 posthatch; 0.025 to 4000 $mg \cdot L^{-1}$ in drinking water) resulted in significant depletion of thyroidal T_4, compared to controls, at 0.05 $mg \cdot L^{-1}$ and all higher concentrations used. In contrast, plasma T_4 was significantly decreased and thyroid gland weights were significantly increased only at ≥500 $mg \cdot L^{-1}$ ammonium perchlorate. Thus, HPT axis responses that lead to release of stored thyroidal hormone (and consequent depletion of thyroidal hormone content) maintain organismal euthyroid status up to high levels of perchlorate exposure at least for a short time scale. Measurements of T_3 in the thyroids and plasma gave essentially the same patterns as for T_4 (McNabb, Larsen, Pooler 2004).

Comparison of the responses of bobwhite chicks to 2, 4, and 8 weeks of exposure to ammonium perchlorate indicated that there is developmental plasticity in the thyroid system that is capable of some adaptation (i.e., compensation) to perchlorate exposure. Thus at 2 weeks of exposure, thyroidal T_4 is significantly decreased at ≥0.05 $mg \cdot L^{-1}$, whereas at 8 weeks of exposure, thyroidal T_4 is significantly decreased only at ≥50 $mg \cdot L^{-1}$ (Figure 5-3). By 8 weeks of exposure, 3 exposure-dependent patterns were apparent:

1) For the lowest concentration range used (0.013 to 5.0 $mg \cdot L^{-1}$ ammonium perchlorate in the drinking water), thyroidal T4 content of treated birds was restored to control levels, that is, there was full compensation for the perchlorate effects.
2) At the intermediate ammonium perchlorate concentrations (50 and 250 $mg \cdot L^{-1}$ in drinking water), there was partial compensation.
3) At the highest concentrations used (500 to 4000 $mg \cdot L^{-1}$ in drinking water), depletion of thyroidal hormone stores continued during the period between 2 and 8 weeks of exposure (Figure 5-4; and see McNabb, Jang, Larsen 2004).

Mallard ducklings

Studies of ducklings drinking ammonium perchlorate solutions from 0.025 to 250 $mg \cdot L^{-1}$ indicated that plasma THs and thyroid gland weights were unaffected but that thyroidal T_4 was decreased at ≥5.0 $mg \cdot L^{-1}$ at 2 weeks of exposure. The pattern of these results was similar to those in quail chicks, but ducklings were less sensitive to ammonium perchlorate than quail. Thus, after a 2-week ammonium perchlorate exposure, the lowest ammonium perchlorate concentration that caused significant thyroidal hormone depletion in ducklings (5.0 $mg \cdot L^{-1}$) was 100× that causing a thyroid effect in quail chicks (0.05 $mg \cdot L^{-1}$). There have been no studies of the effects of ammonium perchlorate exposure for times longer than 2 weeks in mallards (McNabb et al. 2003; McNabb 2004).

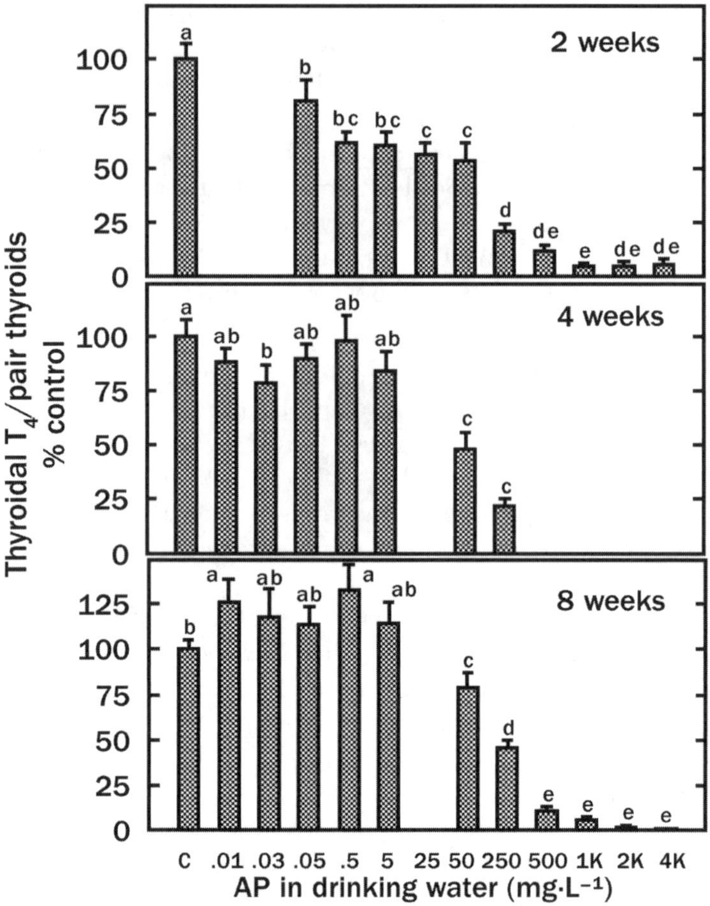

Figure 5-3 Thyroidal T_4 content (ng T_4/pair of thyroid glands) in bobwhite quail chicks after 2, 4, and 8 weeks of exposure to ammonium perchlorate in drinking water. Data are combined from several studies and are expressed as % of control values within each study. Values are means ± standard error. (McNabb, Jang, Larsen 2004. Does thyroid function in developing birds adapt to sustained ammonium perchlorate exposure? Toxicol Sci 82:106–113, by permission of Oxford University Press).

Bobwhite adults

Studies of adult bobwhite quail exposed to perchlorate (0.05 to 8000 $mg \cdot L^{-1}$) for 3, 5, and 8 weeks indicate that adult quail are initially tolerant of higher concentrations of ammonium perchlorate than are chicks. Thus, after 3 weeks of perchlorate exposure, adult quail showed no alterations in their circulating THs or in their thyroid gland weights. Likewise, thyroidal T_4 content was not significantly altered at any of the

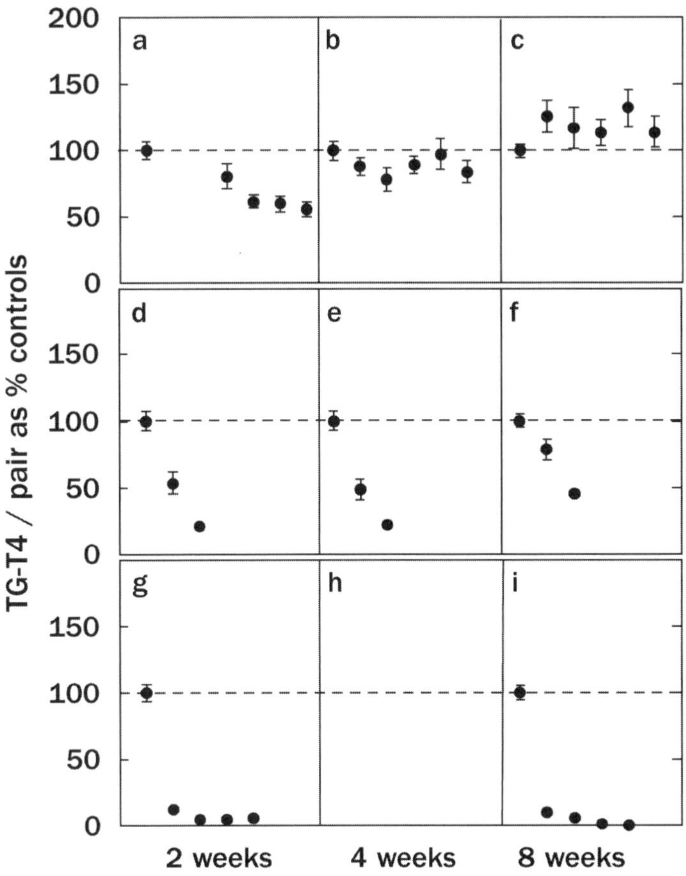

Figure 5-4 Adaptive thyroid responses (of thyroidal T_4 content) in developing bobwhite quail in response to sustained ammonium perchlorate exposure. Ammonium perchlorate concentrations in drinking water in panels a, b, and c are 0, 0.13, 0.25, 0.05, 0.5, and 5.0 $mg \cdot L^{-1}$; in panels d, e, and f, they are 0, 50, and 250 $mg \cdot L^{-1}$; and in panels g and i, they are 500, 1000, 2000, and 4000 $mg \cdot L^{-1}$. (McNabb, Jang, Larsen 2004. Does thyroid function in developing birds adapt to sustained ammonium perchlorate exposure? Toxicol Sci 82:106–113, by permission of Oxford University Press).

ammonium perchlorate exposures used, although there was a strong trend to decreases in thyroidal TH content at ≥250 $mg \cdot L^{-1}$ in the drinking water. In adult quail, the effects of ammonium perchlorate increased with increasing exposure time. By 5 weeks of exposure, thyroidal T_4 content was significantly depleted (by >50%) at ≥250 $mg \cdot L^{-1}$, and by 8 weeks of exposure, thyroidal T_4 was significantly depleted at 0.05 $mg \cdot L^{-1}$. Plasma T_4 concentrations were decreased and thyroid gland weights were increased compared to controls only at ≥4000 $mg \cdot L^{-1}$ after 8 weeks of exposure.

There was no indication that adult quail show adaptive responses to sustained ammonium perchlorate exposure, at least within the time frame of this study (McNabb 2004).

Thyroid histology of laying bobwhite hens is altered by perchlorate exposure, suggesting HPT axis activation (Gentles et al. 2005). In this study, hens were exposed to 1.18, 11.75, and 117.5 $mg \cdot L^{-1}$ ammonium perchlorate in drinking water for 30 days. In hens exposed to 117.5 $mg \cdot L^{-1}$, the height of thyroid follicle cells was increased significantly (a shift from cuboidal to columnar), and the colloid area of follicles was decreased significantly (Figures 5-5 and 5-6). In addition, the glands tended to be hyperplastic. This histological picture is characteristic of TSH-stimulated thyroid glands. Although plasma THs were not measured in this study, it seems likely that the hens were not severely hypothyroid because Japanese quail hens cease laying when plasma THs are decreased in hypothyroidism (Wilson and McNabb 1997). Unfortunately, the exposure times in the study of Gentles et al. (2005) and those of McNabb (2004) differ, so they do not yield comparative information about the sensitivity of histology in relation to measurements of other thyroid variables.

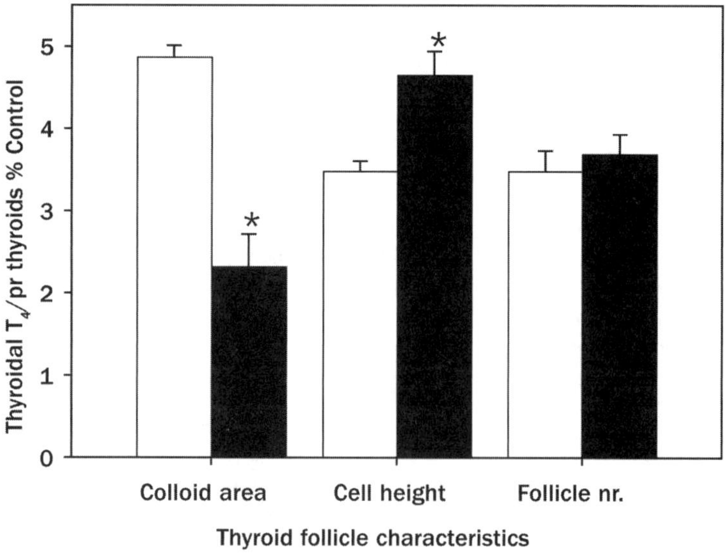

Figure 5-5 The effect of ammonium perchlorate on colloidal area ($\mu m^2 \times 10^5$), epithelial cell height (μm), and number of follicles per thyroid gland section (follicle × 100) in bobwhite quail hens. Open bars, control hens; black bars, hens drinking 117.5 $mg \cdot L^{-1}$ ammonium perchlorate. Asterisks denote significant differences from controls. Values are means ± SE. (Reprinted with permission from Gentles et al. 2005; copyright Society of Environmental Toxicology and Chemistry [SETAC])

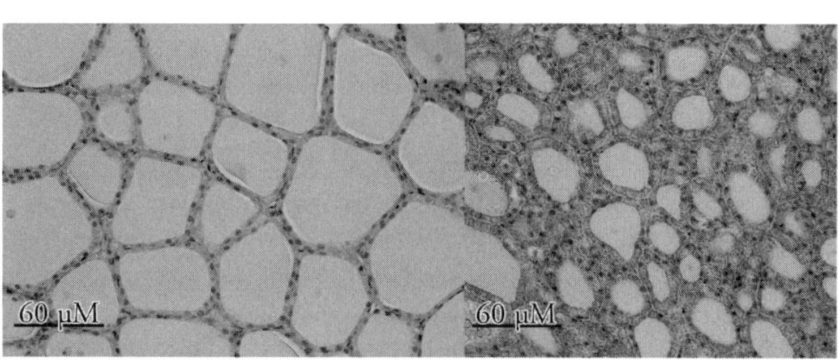

Figure 5-6 Thyroid gland histology in bobwhite quail hens. Left panel: controls; hens drinking distilled water. Right panel: hens drinking 117.5 mg·L^{-1} ammonium perchlorate. (Reprinted with permission from Gentles et al. 2005; copyright Society of Environmental Toxicology and Chemistry [SETAC])

Comparisons of the effects of sodium versus ammonium cations with perchlorate on thyroid function

Although most perchlorate contamination originates from the release of ammonium perchlorate, most environmental perchlorate is thought to exist as sodium perchlorate (Urbansky 1998). Several experiments comparing the effects of equimolar amounts of perchlorate as sodium versus ammonium salts have been done on bobwhite chicks and embryos and on mallard embryos. In addition, some studies have addressed the effects of ammonium ion by administering ammonium chloride. In general, the results show that ammonium and sodium perchlorate have essentially identical effects on avian thyroid function. Surprisingly, ammonium chloride in drinking water seemed to stimulate thyroid function in some experiments (McNabb and Queral-Kirkpatrick 2003; McNabb 2004).

Target organ effects of perchlorate-induced hypothyroidism in developing birds

The only studies of the effects of perchlorate-induced hypothyroidism on TH target tissues in birds have addressed body and skeletal growth. Those studies are described in "Body and skeletal growth," below. However, there are a number of studies that demonstrate TH effects on the development of avian tissues and organs. Thus, these tissues are fruitful areas for investigation of the downstream effects of thyroid disruption resulting from perchlorate (see "Body and skeletal growth," p 112; "Skeletal tissues, muscle and feather development," p 112; "Central nervous system development," p 113; and "Metabolism and thermoregulation," p 114).

Body and skeletal growth

Because THs are required for development (differentiation and growth), it is often suggested that growth-related variables should be useful assessment tools for detecting the effects of alterations in thyroid function. Body weight and femur and tibia length in growing quail have been evaluated to determine if target organ responses to thyroid alterations are affected by ammonium perchlorate exposure. In precocial bobwhite chicks exposed to ammonium perchlorate, these growth-related variables were less sensitive indicators of ammonium perchlorate exposure than were any of the thyroid variables measured. For example, after 8 weeks, there were no effects of ammonium perchlorate on body weight, and femur and tibia length were significantly decreased only at the highest ammonium perchlorate concentration used (4000 $mg \cdot L^{-1}$; McNabb, Larsen, Pooler 2004). Bobwhite chicks drinking ≥500 $mg \cdot L^{-1}$ ammonium perchlorate develop organismal hypothyroidism by 2 weeks of exposure and experience sustained hypothyroidism with exposure to these concentrations up to at least 8 weeks, the longest period tested (McNabb, Jang, Larsen 2004). These results suggest that severe hypothyroidism, for sustained periods of time, is necessary before posthatch growth patterns are affected in this precocial species. Likewise, femur and tibia length were not altered in 21-d bobwhite quail embryos by in ovo ammonium perchlorate exposures from 50 to 150 $\mu g \cdot g^{-1}$ of egg (McNabb 2004). This result is not surprising because quail embryos exposed to this range of ammonium perchlorate concentrations maintain euthyroid levels of circulating THs despite having decreased thyroidal T_4 stores (see "Bobwhite chicks," p 106). These results are in contrast to the decreases in hindlimb growth observed in a study of chicken embryos made hypothyroid by in ovo PCB treatment (Gould et al. 1999).

Recently, an investigation in nestling zebra finches (*Taeniopygia guttata*) dosed orally with 0, 10, 100, or 1000 $mg \cdot kg^{-1}$ sodium perchlorate on post-hatch days 3 to 14, has demonstrated perchlorate effects on posthatching growth. In this altricial species, there were significant decreases in body mass and tibiotarsal length in birds from the 100 and 1000 $mg \cdot kg^{-1}$ dose groups. Thyroid studies in these birds are in progress (Rainwater et al. 2004).

Skeletal tissues, muscle and feather development

Avian skeletal, muscular, feather, and nervous system differentiation and growth are influenced by THs directly or in concert with other growth factors. The influence of a hypothyroid state on bone development has been demonstrated in mammals, and more limited studies suggest TH effects on these tissues are the same in birds (McNabb 2000). Studies of the roles of THs in the differentiation of pelvic cartilages in chicken embryos have illustrated how THs interact with growth factors in cartilage-to-bone developmental transitions. In brief, insulin-like growth factors are involved in the initiation of cartilage growth in pelvic cartilages, but T_3 stimulates the maturation of chondrocytes (Burch and Lebovitz 1982; Burch and Van Wyk 1987). Pelvic cartilage growth and alkaline phosphatase activity (which is indicative of this cartilage differentiation) have been shown to reflect the level of maternal T_4 deposited in quail eggs when hens are treated with T_4 (Wilson and McNabb 1997). Hypothyroidism in developing barnacle geese (*Branta leucopsis*) also led to alterations in skeletal development (Deaton et al. 1998).

THs are required for the differentiation of muscle cells in birds, especially for the transitions from neonatal forms of myosin to the adult heavy chain, fast myosin (King et al. 1987). This transition is blocked at least temporarily by treatment with the goitrogen methimazole (Maruyama et al. 1995). Thyroid status also affects the relative muscle mass and the characteristics of different muscles (e.g., cardiac versus pectoralis muscle; Deaton et al. 1998; Bishop et al. 2000) as well as the proportion of fast and slow twitch fibers in developing avian muscle (Dainat et al. 1991). Differences in muscle fiber types in turn are related to differences in the number of thyroid receptors in different muscles (Dainat et al. 1984, 1986) and influence the aerobic capacity of locomotor muscles (Bishop et al. 1995). Differences in thyroid status have differential effects on thyroid receptor expression in different muscles (Bishop et al. 2000).

Hypothyroidism in adult chickens is known to alter the pattern of molting and feather characteristics and has long been utilized to manipulate molt cycles in poultry production (Decuypere and Verheyen 1986). Feather growth in developing barnacle geese also was altered by hypothyroidism (Deaton et al. 1998).

Central nervous system development

Perhaps the most intriguing system that may be affected by a hypothyroid state is the development of the nervous system. THs are required for normal brain development. During both gestational and postnatal periods in mammals, THs play an important role in neuronal proliferation, migration, synaptogenesis, and myelination (Howdeshell 2002; Zoeller et al. 2002; Chapter 3). In adult birds, a hyperthyroid state leads to increased levels of focused neuronal cell death (Tekumalla et al. 2002), indicating that THs play an important role in nerve cell turnover in the brain, a process that is best known in birds and whose potential is of great interest in mammals. Because the death and turnover of brain cells in birds is closely tied to TH levels, an understanding of how a perchlorate-induced hypothyroid state influences neuron development, in both nestlings and adults (i.e., annual song nucleus growth and regression) would be useful in assessing the potential for perchlorate developmental neurotoxicity. Alterations in the patterns of seasonal brain development in birds may lead to behavioral changes that affect survival and reproduction. The development of the song nucleus in the zebra finch has been used as a target to assess the impacts of hormonal perturbances on neurodevelopment (Quaglino et al 2002). Preliminary studies in nestling zebra finches are underway to assess the potential for perchlorate neurodevelopmental effects (Rainwater et al. 2004).

Metabolism and thermoregulation

In addition to growth and development, metabolic effects (on oxygen consumption and glycogen utilization) of thyroid function are seen in the increased levels of THs associated with cold temperatures and the maintenance of constant body temperatures in birds (Freeman 1970, 1971). Studies of organophosphate insecticides in birds have established that loss of thermoregulatory capacity can affect survival in birds exposed to cold temperatures, having the greatest potential effect in species with small body sizes and high metabolic rates (Martin and Soloman 1991; Grue et al. 1997). This loss of thermoregulatory capacity is of concern particularly during the perihatch period in precocial birds and during the nestling period in altricial birds, when increased circulating THs are associated with the initiation of thermoregulation (see "Modes of development," p 101). Assessment of the effects of perchlorate on thermoregulation in birds, particularly species with high metabolic rates (under fed and unfed conditions), and in nestlings during the perihatch period would provide insight into the role that perchlorate-induced hypothyroidism plays in disrupting metabolic homeostasis and thermoregulation.

Ecotoxicology of Perchlorate Exposure in Birds

Determinants of ecotoxicity in birds

The ecotoxicological risk of perchlorate in birds depends, as with other species, on the toxicity of the compound and the potential for exposure levels that cause adverse effects. The previous sections detail what is known about the effects of perchlorate exposure in model avian species, demonstrating that there is variability in sensitivity between species, that chicks initially appear more sensitive than adults, but that chicks (but not adults) can show some adaptation to perchlorate with continued low- to moderate-level exposures. With these current data on perchlorate effects on birds, it is useful to consider whether wild birds are exposed to perchlorate levels that might be consistent with those that alter thyroid function and that may have adverse effects on target organs or systems.

As with most wildlife species, exposure of birds to any contaminant is most likely to occur through the ingestion of contaminated food, water, and soil. Thus, a good knowledge of environmental perchlorate contamination and of the physiology and natural history of the avian species under consideration is necessary. Knowledge of the composition of the diet, the strategy for obtaining food, and the feeding dynamics is critical to predicting the likelihood of exposure in each species. In addition, the physical and chemical characteristics of the contaminant, the environmental distribution pattern of the contaminant in the environment, and the bioavailability of the contaminant all play important roles in avian contaminant exposure and uptake.

Birds have a variety of species-specific differences in diet composition and foraging strategies. Generally, bird species can be grouped into several categories that describe their dietary preferences and foraging strategies, for example, granivores, frugivores, carnivores (predatory birds), piscivores, and omnivores (scavengers). Of course, these categories describe diet composition in its broadest terms, and specific diet composition varies considerably among species and individuals based on seasonal and site-specific availability of food and on species-specific preferences for particular diet items. Also, foraging strategy, or the mechanics of when and where food is gathered, varies among species both within and among diet categories. Migration, size of the foraging area, specific habitats used for foraging, life stage, and season all determine where and when specific foods are obtained, and consequently affect the potential for exposure to contaminants. For example, nestling birds are often fed only insects, although adults of the same species may be granivorous at all other times.

Potentially, any species of bird could come into contact with perchlorate through the ingestion of contaminated media. However, the high solubility of perchlorate in water, and thereby its known association with water and adjacent terrestrial habitats, suggests the most likely avian receptors will include birds that consume contaminated free-standing water or food. For example, obvious candidates for perchlorate exposure include shorebirds and waterfowl that forage in wetland areas. These birds may skim food from the surface of the water, probe for food in sediments, or pick up food lying on the sediment surface, thereby gaining exposure to perchlorate in both their food and water. Perchlorate has been shown to readily move into plants and accumulate in leaves, seeds, and fruit (Smith et al. 2001; Tan et al. 2004; Jackson et al. 2005), suggesting that birds consuming seeds and fruits would likely be exposed if they are foraging in contaminated areas. Examples of perchlorate accumulation in seeds include commercially grown wheat seed (*Triticum* sp.; 1.5 to 9.5 mg $ClO_4^- \cdot kg^{-1}$ dw; Jackson et al. 2005) and smartweed seeds (*Polygonum* sp.; 40.6 mg $ClO_4^- \cdot kg^{-1}$ dw; Tan et al. 2004) collected along a contaminated stream. More extreme examples include seeds from crabgrass (*Digitaria* sp.; 1880 mg $ClO_4^- \cdot kg^{-1}$ dw) and goldenrod (*Solidago* sp.; 184 mg $ClO_4^- \cdot kg^{-1}$ dw) plants collected from a contaminated site in Texas (Smith et al. 2001). Accumulation of perchlorate in fruit likely to be consumed by birds has also been documented, with mean concentrations in mulberry fruit (*Broussonetia papyrifera*) of 0.47 mg $ClO_4^- \cdot kg^{-1}$ dw (Tan et al. 2004). In contrast, perchlorate was not detected in chinaberry fruit (*Malia azedarach*) collected from trees along the same stream where the mulberry fruit was collected. Perchlorate was detected in other, less palatable, plant parts in these studies (e.g., stems and leaves). Although consumption of stems and leaves by birds would be unlikely or minor, these plant parts may serve as a conduit for trophic transfer of perchlorate to birds through consumption of herbivorous invertebrates (e.g., Lepidoptera larvae). Perchlorate can accumulate in invertebrates and vertebrates (Smith et al. 2001), providing a route of exposure for insectivores, raptors, scavengers, and piscivores. However, in all cases, the potential for exposure will likely be restricted to birds that forage on specific plants or animals associated with perchlorate-contaminated water or soil.

Perhaps the greatest potential for exposure of birds to relatively high levels of perchlorate is through consumption of contaminated water. Birds vary widely in the magnitude of their water turnover. Some birds do consume standing water regularly, although other species obtain their water primarily from their food or through metabolic production (Bartholomew 1982). As a group, granivorous birds are likely to consume the most free-standing water because their diets are naturally low in moisture (Bartholomew and Cade 1963; Gill 1995). Birds that are strong fliers can travel long distances to reach open water sources and often have high water turnover. In contrast, ground-dwelling, weak fliers are dependent on local water sources (review, Bartholomew and Cade 1963). Birds often will return to the same source of water on a daily basis, especially in arid environments, thereby greatly increasing their exposure if the water source is contaminated.

Local water sources such as highly contaminated holding ponds or small ponds contaminated by perchlorate in the soil are likely to have the highest perchlorate concentrations; large bodies of water are likely to have much lower concentrations of perchlorate. Concentrations of perchlorate in standing water bodies can vary tremendously. Mean perchlorate levels in water from the Las Vegas Wash in Nevada, USA, ranged from 0.15 to 1.04 μg $ClO_4^- \cdot mL^{-1}$ along approximately 2 miles of stream (Smith et al. 2004). Water samples collected from the same site previously reported by Smith et al. (2001) contained perchlorate at concentrations that ranged from nondetectable to 85 $ng \cdot mL^{-1}$ in running streams. Although this is an extreme example, these authors also reported perchlorate concentrations of over 30 $μg \cdot mL^{-1}$ in a pond used to hold ground water prior to release into a neighboring stream. Migratory birds that are exposed to perchlorate for only part of the year may be able to recover from perchlorate effects during the non-exposure season. Other studies have found groundwater perchlorate concentrations up to 630 $mg \cdot L^{-1}$ near perchlorate manufacturing sites (Urbansky 1998).

Relevance of potential exposure levels

The goal of this chapter is to document the avian ecotoxicology of perchlorate, and it is, thus, beyond the scope of this effort to develop a risk assessment for perchlorate in wild birds, an exercise addressed in Chapter 10. The lack of measured exposure data for birds inhabiting perchlorate-contaminated environments suggests the need for a preliminary evaluation of the relevance of perchlorate levels found in avian food and water sources. Of the 2 species tested, more is known about how bobwhites respond to perchlorate, and bobwhites are more sensitive to perchlorate than mallards, so they will be considered in this assessment.

Perchlorate concentrations of 0.085, 0.15, 1.0, and 30 mg $ClO_4^- \cdot L^{-1}$ in natural bodies of water were noted in the examples of the previous section. At all of these levels, prolonged exposure of adult bobwhites would lead to depression of thyroidal T_4 content and, at the highest water level, might lead to depressed thyroidal T_4 content in embryos and hatchlings from water-consuming hens. Though bobwhite chicks also would suffer depressed thyroidal T_4 content initially, current evidence suggests that thyroid function in chicks has sufficient developmental plasticity to adapt to these levels of perchlorate at least for exposures of 8 weeks. Based on these perchlorate water-exposure levels, it does not appear likely that there would be organismal hypothyroidism (plasma T_4 depression), sustained HPT axis responses (thyroid gland weight increases), or target organ effects resulting from alterations in thyroid function (e.g., reduction of bone length). However, it should be noted that no sensitive measures of target organ effects of altered thyroid function have been developed for birds.

Food consumption rates in captive bobwhite quail range from 72 to 93 g food · kg^{-1} body weight per day, depending on the season (Koerth and Guthery 1990). Using a value of 100 g · $kg^{-1} \cdot d^{-1}$ for simplicity and a worst-case scenario, the examples of perchlorate ion in avian food items noted in "Determinants of ecotoxicity in birds" (p 114) would yield the following daily perchlorate exposures if they were the only food items consumed by bobwhites:

- mulberry seed; 0.05 mg ClO_4^- · kg^{-1} body weight per day
- wheat seed; 0.15 to 0.95 mg ClO_4^- · kg^{-1} body weight per day
- smartweed seed; 4.1 mg ClO_4^- · kg^{-1} body weight per day
- goldenrod seed; 18.4 mg ClO_4^- · kg^{-1} body weight per day
- crabgrass seed; 188 mg ClO_4^- · kg^{-1} body weight per day

Estimated doses for birds drinking perchlorate solutions in laboratory studies can be calculated using published drinking rates (quail, Koerth and Guthery 1990; mallards, USEPA 1993) and allow comparison with potential perchlorate intake in food. For example, these calculations suggest that hens consuming a goldenrod seed diet would have altered thyroid histology and would deposit perchlorate in their eggs at concentrations that produce decreased thyroidal T_4 in embryos and hatchlings from those eggs. Any of the diets listed above would be expected to cause decreased thyroidal T_4 content in adult birds by 8 weeks of exposure. However, none of these diets would be expected to cause decreases in growth of hindlimbs in chicks, at least for exposure times up to 8 weeks, but growth effects are a very insensitive indicator of target organ effects of thyroid deficiency.

Most of the daily exposure levels for diets consisting of the plant materials listed fall within the range of those perchlorate concentrations that cause some thyroid effects. Although the time to effects would vary based on concentrations in the different seeds and the life stage exposed, these time effects could cause alterations in thyroid function. Consumption of perchlorate-contaminated water and incidental consumption of perchlorate-contaminated soils with these food items would further increase the dose received by animals living in contaminated environments.

These findings are estimates that must be considered very preliminary. However, they suggest that under worst-case conditions, birds can obtain perchlorate doses that lead to endocrine disruption and potential target organ effects. To better refine these exposure–response estimates, uncertainties must be considered when laboratory study outcomes are compared with field exposure scenarios. Important uncertainties relevant to the predicted effects noted above include the following:

- Proportion of diet and water contaminated. Because of the heterogeneous nature of contaminant occurrence and the natural movements of birds in the wild state, not all food and water a wild animal consumes is contaminated equally. The relative importance of these 2 avenues of perchlorate intake in any bird species would need to be determined.
- Perchlorate bioavailability. In laboratory studies to date, perchlorate was incorporated into drinking water, an exposure route likely leading to high bioavailability because of its solubilized state. The comparative bioavailability of perchlorate in avian food items or in ingested soil is not known for birds. Similar bioavailability to drinking water was assumed in our estimates.
- Relevance of thyroidal T_4 content versus plasma TH reductions as indicators of thyroid disruption. Reduction of thyroidal T_4 content reflects altered thyroid function (i.e., an early phase of disruption of an endocrine system) but not an adverse effect on target tissues. Substantial depletion of thyroidal T_4 content is required before plasma T_4 is reduced, and sustained exposure to decreased plasma T_4 concentrations is required before target tissue alterations are measurable (at least for variables such as hindlimb growth). It remains to be resolved whether altered endocrine function should be considered indicative of contaminant effects or whether only target organ alterations will be considered as endpoints.
- Adaptation to perchlorate in chicks. Bobwhite chicks, though not adults, have been shown to adapt to sustained perchlorate exposure. A better understanding of this unexpected effect is needed to evaluate perchlorate effects during avian development.
- Cross-species extrapolation. In the examples above, northern bobwhite and mallard ducks demonstrated substantial differences in sensitivity, with ducks having a greater resistance to the effects of perchlorate than quail. The applicability of the findings of these 2 precocial species to other avian species, especially those with altricial development, is unknown. Studies under way using the zebra finch will add to the diversity of avian species studied, providing plasma and thyroidal TH levels, thyroid and brain histopathology, and growth and behavior data for further comparison (Rainwater et al. 2004).

Conclusions

As in other vertebrates, thyroid function in birds is decreased by perchlorate exposure. Recent studies of ammonium perchlorate effects in bobwhite quail and mallard ducks have shown that the most sensitive indicator of alterations in thyroid function is thyroidal TH content. In contrast, measurements of circulating THs (indicative of organismal thyroid status) and thyroid gland weights (indicative of HPT axis activation) are much less sensitive.

Recent studies show that hens deposit perchlorate in their eggs and that egg perchlorate affects embryonic thyroid function, at least in the precocial species that have been investigated to date (bobwhite quail and mallard ducks). These studies show that despite the effects of in ovo perchlorate exposure, embryos maintain euthyroid status, although their thyroidal TH stores are typically depleted at some perchlorate exposures compared to controls. Mallard embryos were less sensitive to perchlorate than were bobwhite embryos. Presumably, the eggs of perchlorate-exposed hens also will be deficient in two other regards that are consequences of perchlorate effects on the hens, namely the deposition of egg I^- and of maternal thyroid hormones. However, to date, these egg components have not been measured in the eggs of perchlorate-treated hens nor have embryos developing in eggs from perchlorate-treated hens been studied. Laboratory studies of in ovo perchlorate exposure suggest that embryonic and hatchling thyroidal T_4 content is significantly decreased at ~1/4 the perchlorate concentration deposited in eggs by hens drinking 117.5 $mg \cdot L^{-1}$ ammonium perchlorate.

Dose–response studies (2 weeks of exposure) in chicks of 2 precocial species indicate that thyroidal T_4 content is significantly decreased compared to controls in bobwhite quail chicks drinking ≥0.05 $mg \cdot L^{-1}$ ammonium perchlorate, whereas in mallard ducklings, thyroidal T_4 content was significantly decreased only at ≥5.0 $mg \cdot L^{-1}$ ammonium perchlorate. Thus, mallard ducklings appear to be considerably less sensitive to perchlorate than are bobwhite quail chicks. Although it is commonly assumed that all developing systems are more susceptible to disruption than adult systems, bobwhite quail chicks appear to show adaptation to some levels of sustained ammonium perchlorate exposure.

Bobwhite adults tolerate higher ammonium perchlorate exposures than do chicks, they develop thyroid effects more slowly than do chicks, and they show no indication of adaptive responses, at least up to 8 weeks of extremely high ammonium perchlorate exposures. Comparisons of ammonium versus sodium perchlorate (equimolar perchlorate ion) in drinking water suggest that different cations combined with the perchlorate anion do not differentially alter the responses of quail chicks and embryos or mallard embryos to perchlorate.

Very little is known about the downstream effects of perchlorate-induced hypothyroidism on target organs or tissues that are responsive to THs. To date, only body and skeletal growth endpoints have been assessed. Both body weight and hindlimb growth (femur and tibia length) are very insensitive indicators and are largely unaffected in birds exposed to high concentrations of ammonium perchlorate for up to 8 weeks and showing overt hypothyroidism as indicated by markedly depressed circulating TH concentrations. Downstream effects on the central nervous system development or the development of skeletal, muscle, or integument and feathers seem more promising than endpoints that have been used to date.

Future Research Needs

- Additional studies are needed to define the measurement protocols that should be used to establish how thyroid disruption or its downstream effects should be evaluated in birds. Specifically, altered thyroid histology should be compared with measurements of thyroidal TH content because these 2 thyroid variables appear to be the most sensitive to thyroid disruption.
- Studies of avian food items and water sources indicate that the levels of exposure required for the thyroid effects described above occur in some situations where environmental perchlorate contamination is present. The occurrence and consequences of a hypothyroid state on bird development, health, survival, and reproduction need to be studied in the context of environmentally relevant exposures.
- All studies to date have been on precocial birds. Comparable studies need to be done on altricial birds because of the differences in the patterns of thyroid, organ system, and thermoregulatory ontogeny in these 2 developmental modes.
- Additional studies at low, environmentally relevant perchlorate concentrations and for longer time periods are needed to fully assess the nature of exposures that cause organismal hypothyroidism and downstream target organ effects.
- Studies of perchlorate, in the context of the background literature on hypothyroid affects on developing avian systems (specifically,

nervous, skeletal, and muscular tissues as well as feather molting), need further investigation.

- Neurodevelopmental effects influencing behavior, survival. and reproduction in nestlings and in adults need to be investigated in birds. Current research on the effects of THs on central nervous system development in mammals, as well as recent studies of TH effects on the plasticity of neuronal cell turnover in the song nucleus of the avian brain, suggests the need for investigations in this area.
- Because THs are important in the control of metabolism, the effects of perchlorate-induced hypothyroidism on the development of thermoregulation should be investigated in both altricial and precocial birds. These studies may be especially fruitful in very small birds with very high metabolic rates. Hypothyroidism may affect thermoregulatory development in nestling altricial young directly and may have indirect effects through the high energetic demands on parental care of nestlings.
- The interactions between environmental iodine availability and perchlorate effects in birds need to be investigated. High iodine availability is known to lessen the effects of perchlorate in humans and laboratory mammals (see Chapter 3), but there is no available information about this interaction in birds.

References

Bartholomew GA. 1982. Body temperature and energy metabolism. In: Gordon MS, editor. Animal physiology: principles and adaptations, 4th ed. New York (NY): Macmillan. p 333–406.

Bartholomew GA, Cade TJ. 1963. The water economy of land birds. Auk 80:504–539.

Bishop CM, Butler PJ, Atkinson NM. 1995. The effect of elevated levels of thyroxine on the aerobic capacity of locomotor muscles of the tufted duck, *Aythya fuligula.* J Comp Physiol 164B:618–621.

Bishop CM, McCabe CJ, Gittoes NJL, Butler PJ, Franklyn JA. 2000. Tissue-specific regulation of thyroid hormone receptor mRNA isoforms and target gene proteins in domestic ducks. J Endocrinol 165:607–615.

Brown TR, Sundick RS, Dhar A, Sheth D, Bagchi N. 1991. Uptake and metabolism of iodine is crucial for the development of thyroiditis in Obese Strain chickens. J Clin Investig 88:106–111.

Burch WM, Lebovitz HE. 1982. Triiodothyronine stimulation of in vitro growth and maturation of embryonic chick cartilage. Endocrinology 11:462–468.

Burch WM, Van Wyk JJ. 1987. Triiodothyronine stimulates cartilage growth and maturation by different mechanisms. Am J Physiol 252:E176–E182.

Dainat J, Bressot C, Bacou F, Rebiere A, Vigneron P. 1984. Perinatal age and sex variations of the triiodothyronine nuclear receptors in the chick pectoralis major muscle. Mol Cell Endocrinol 35:215–220.

Dainat J, Bressot C, Rebiere A, Vigneron P. 1986. Ontogenesis of triiodothyronine nuclear receptors in three skeletal muscles in male and female chicks. Gen Comp Endocrinol 62:479–484.

Dainat J, Saleh L, Bressot C, Marger L, Bacou F, Vigneron P. 1991. Effects of thyroid state alterations in-ovo on the plasma levels of THs and on the population of fibers in the plantaris muscle of male and female chickens. Reprod Nutr Dev 31:703–716.

Deaton KE, Bishop CM, Butler PJ. 1998. Tissue-specific effects of hypothyroidism on postnatal muscle development in the barnacle goose. J Exp Biol 201:827–36

Decuypere E, Verheyen G. 1986. Physiological basis of induced moulting and tissue regeneration in fowls. World's Poult Sci J 42:56–66.

Delange FM, Ermans A-M. 1996. Iodine deficiency. In Braverman LE, Utiger RD, editors. Werner and Ingbar's the thyroid, 7th ed. Philadelphia (PA): Lippincott-Raven. p 296–316.

Freeman BM. 1970. Thermoregulatory mechanisms of the neonate fowl. Comp Biochem Physiol 33:219–230.

Freeman BM. 1971. Impaired thermoregulation in the thiouracil-treated fowl. Comp Biochem Physiol 40A:553–555.

Gentles A, Surles J, Smith EE. 2005. Evaluation of adult quail and egg production following exposure to perchlorate-treated water. Environ Toxicol Chem 24:1930–1934.

Gill FB. 1995. Ornithology. 2nd ed. New York (NY): WH Freeman. p 763.

Gould JC, Cooper KR, Scanes CG. 1999. Effects of polychlorinated biphenyls on thyroid hormones and liver Type I monodeiodinase in the chick embryo. Ecotoxicol Environ Saf 43:195–203.

Grue CE, Gibert PL, Seeley ME. 1997. Neurophysiological and behavioral changes in non-target wildlife exposed to organophosphate and carbamate pesticides: Thermoregulation, food consumption and reproduction. Am Zool 37:369–388.

Howdeshell KL. 2002. A model of the development of the brain as a construct of the thyroid system. Environ Health Perspect 110 Suppl 3:337–348.

Jackson WA, Joseph P, Laxman P, Tan K, Smith PN, Yu L, Anderson TA. 2005. Perchlorate accumulation in forage and edible vegetation. J Agric Food Chem 53:369–373.

King DB, Bair WE, Jacaruso RB. 1987. Thyroidal influence on nuclear accumulation and DNA replication in skeletal muscles of young chickens. J Exp Zool Suppl 1:291–298.

Koerth NE, Guthery FS. 1990. Water requirements of captive northern bobwhites under subtropical seasons. J Wildl Manag 54:667–672

Martin PA, Soloman KR. 1991. Acute carbofuran exposure and cold stress: Interactive effects in mallard ducklings. Pestic Biochem Physiol 40:117–127.

Maruyama K, Kanemaki N, May JD. 1995. Thyroid hormones on embryo development and appearance of myosin heavy chain isoforms in turkeys (*Meleagris gallopavo*). Comp Biochem Physiol 112C:109–117.

McNabb FMA. 2000. Thyroids. In: Whittow GC, editor. Sturkie's avian physiology, 5th ed. New York (NY): Academic Pr. p 461–471.

McNabb FMA. 2004. The effects of perchlorate on developing and adult birds. Final report. Washington (DC): Strategic Environmental Research Development Program.

McNabb FMA. 2005. Biomarkers for the assessment of avian thyroid disruption by chemical contaminants. Poult Avian Biol Rev 16:3–10.

McNabb FMA, Blackman JR, Cherry JA. 1985. The effects of different maternal dietary iodine concentrations on Japanese quail. I. Thyroid status of hens. Dom Anim Endocrinol 2:25–34.

McNabb FMA, Jang DA, Larsen CT. 2004. Does thyroid function in developing birds adapt to sustained ammonium perchlorate exposure? Toxicol Sci 82:106–113.

McNabb FMA, King DB. 1993. Thyroid hormone effects on growth, development and metabolism. In: Schreibman MP, Scanes CG, Pang PKT, editors. The endocrinology of growth, development, and metabolism of vertebrates. New York (NY): Academic Pr. p 393–417.

McNabb FMA, Larsen CT, Pooler PS. 2004. Ammonium perchlorate effects on thyroid function and growth in bobwhite quail chicks. Environ Toxicol Chem 23:997–1003.

McNabb FMA, Maher RD, Jones JE, Jang DA. 2003. Ammonium perchlorate effects on developing and adult thyroid function in bobwhite quail and mallard ducks. Soc Integr Comp Biol Abstr. 43.1, p 247.

McNabb FMA, Olson JM. 1996. Development of thermoregulation and its hormonal control in precocial and altricial birds. Poult Avian Biol Rev 7:111–125.

McNabb FMA, Queral-Kirkpatrick LT. 2003. Perchlorate exposure level, exposure time and perchlorate-associated cation effects in birds. In: Abstract Book, SETAC 24th Annual Meeting in North America; 2003 Nov 16–20; Austin, TX. Pensacola (FL): Society of Environmental Toxicology and Chemistry (SETAC). p 57. Abstract nr 250.

McNabb FMA, Scanes CG, Zeman M. 1998. The endocrine system. In: Starck JM, Ricklefs RE, editors. Avian growth and development: evolution within the altricial precocial spectrum. Oxford (UK): Oxford Univ Pr. p 174–202.

McNabb FMA, Schreck CB, Tyler C, Thomas P, Kramer VJ, Specker JL, Mayes MA, Selcer KW. 1999. Basic physiology. In: Di Giulio RT, Tillitt DE, editors. Reproductive and developmental effects of contaminants in oviparous vertebrates: mechanisms, ecological consequences and assessments of risk. Pensacola (FL): Society of Environmental Toxicology and Chemistry. p 113–223.

McNichols MJ, McNabb FMA. 1988. Development of thyroid function and its pituitary control in embryonic and hatchling precocial Japanese quail and altricial Ring doves. Gen Comp Endocrinol 69:109–118.

Morreale de Escobar G, Pastor R, Obregon MJ, Escobar del Rey F. 1985. Effects of maternal hypothyroidism on the weight and thyroid hormone content of rat embryonic tissues before and after onset of fetal thyroid function. Endocrinology 117:1890–1900.

Peña HG, Barth KM, Kessler WV, Plumlee MP, Christian JE. 1966. Alteration of the incorporation of Iodine-131 in the chicken egg. Poult Sci 46:426–429.

Peña HG, Kessler WV, Christian JE, Cline TR, Plumlee MP. 1975. A comparative study of iodine and potassium perchlorate metabolism in the laying hen. I. Incorporation of I^{-}131 in chicken ova. Poult Sci 54:170–183.

Peña HG, Kessler WV, Christian JE, Cline TR, Plumlee MP 1976. A comparative study of iodine and potassium perchlorate metabolism in the laying hen. 2. Uptake, distribution, and excretion of potassium perchlorate. Poult Sci 55:188–201.

Quaglino AE, Craig-veit CB, Viant MR, Erichsenn AL, Fry DM, Millam JR. 2002. Oral estrogen masculinizes female zebra finch song system. Horm Behav 41:236–241.

Rainwater TR, Wood MB, Millam JR, Hooper MJ. 2004. Effects of sodium perchlorate on growth and development in zebra finch chicks [abstract]. Society of Environmental Toxicology and Chemistry World Congress and 25th Annual Meeting; 2004 Nov 14–18; Portland, OR.

Smith PN, Theodorakis CW, Anderson TA, Kendall RJ. 2001. Preliminary assessment of perchlorate in ecological receptors at the Longhorn Army Ammunition Plant (LHAAP), Karnack, Texas. Ecotoxicology 10:305–313.

Smith PN, Yu L, McMurry ST, Anderson TA. 2004. Perchlorate in water, soil, vegetation, and rodents collected from the Las Vegas Wash, Nevada, USA. Environ Pollut 132:121–127.

Tan K, Anderson TA, Jones MW, Smith PN, Jackson WA. 2004. Accumulation of perchlorate in aquatic and terrestrial plants at a field scale. J Environ Qual 33:1638–1646.

Taurog A. 1996. Hormone synthesis: thyroid iodine metabolism. In: Braverman LE, Utiger RD, editors. Werner and Ingbar's the thyroid, 7th ed. Philadelphia (PA): Lippincott-Raven. p 47–81.

Tekumalla PK, Tontonoz M, Hesla MA, Kirn JR. 2002. Effects of excess thyroid hormone on cell death, cell proliferation, and new neuron incorporation in the adult zebra finch telencephalon. J Neurobiol 51:323–341.

Thommes RC. 1987. Ontogenesis of thyroid function and regulation in the developing chick embryo. J Exp Zool Suppl 1:273–279.

Urbansky ET. 1998. Perchlorate chemistry: Implications for analysis and remediation. Bioremed J 2:81–95.

[USEPA] US Environmental Protection Agency. 1993. Wildlife exposure factors handbook. Washington (DC): USEPA Office of Research and Development (8603). EPA/600/R-93/187.

Wilson CM, McNabb FMA. 1997. Maternal thyroid hormones in Japanese quail eggs and their influence on embryonic development. Gen Comp Endocrinol 107:153–165.

Zoeller RT, Dowling ALS, Herzig CTA, Iannacone EA, Gauger KJ, Bansal R. 2002. Thyroid hormone, brain development, and the environment. Environ Health Perspect 110:355–361.

six

Effects of Perchlorate in Amphibians

James A Carr, Christopher Theodorakis

Ecological Importance of Amphibians

There are 3 taxonomic orders of extant amphibians within the class amphibia:

1) order Anura, which includes all tailless amphibians (frogs and toads);
2) order Caudata (urodeles), which includes the newts and salamanders; and
3) order Gymnophiona (Apoda), which is composed of the caecilians.

These 3 orders make up more than 6000 species distributed worldwide (UNEP–WCMC Species Database 2006). Approximately 283 species can be found in the United States, of which 152 species are endemic (World Resources Institute, Earthtrends 2004).

Amphibians live in a diverse array of habitats, from equatorial rainforests to Alaska, and not all species spend part of their life history on both water and land. For example, some frogs (family Pipidae) and salamanders (family Cryptobranchidae) are highly adapted for an aquatic existence and spend little if any time on land. In contrast, other frog (family Leptodactylidae) and salamander (family Plethodontidae, the lungless salamanders) species do not normally visit water as part of their life history strategies (Feder 1992). Likewise, amphibian modes of development and reproduction are diverse. Although many frog species go through metamorphosis as part of their postembryonic development, there are frog species (*Eleutherodactylus coqui*) that exhibit direct development with no larval stage. Although external fertilization takes place in most amphibian species, caecilian species employ

Perchlorate Ecotoxicology. Ronald J Kendall and Philip N Smith, editors.

ISBN-10: 1-880611-90-2 ISBN-13: 978-1-880611-90-6

internal fertilization, and three-fourths of the species within the order are viviparous or live-bearing. Given this diversity of life history patterns, it is important to use care when attempting to extrapolate from data collected on a single species to all amphibians.

Potential Role of Contaminants in Amphibian Declines

Several lines of evidence support anecdotal observations that frog populations are declining globally. Recent data have quantified large-scale declines in amphibian populations inhabiting North America as well as, in some cases, population declines in genetically isolated species with an extremely limited geographic distribution (Houlahan et al. 2000). Urban encroachment and consequent loss of ecological corridors for migration have resulted in the reduction of some North American species to a single natural population, such as the case with the Wyoming (*Bufo baxteri*) and Houston (*Bufo houstonensis*) toads (USFWS 1984, 2001). In other cases, species with historically wide distribution within the Sierra Nevada, such as the California red-legged frog (CRLF; *Rana aurora draytonii*) or the Cascades frog (*Rana cascadae*), have incurred significant loss (70% in the case of the CRLF) in their geographic distribution (Fellers and Drost 1993). In the most dramatic cases, entire species have become extinct. Two species of *Rheobatrachus*, the gastric brooding frog, have not been observed since the 1980s (Ingram and Donald 1993; Hines et al. 1999).

Before we can consider the potential effects of perchlorate exposure on amphibian declines, it is important to look at the relative role that contaminant exposure may play in this phenomenon. Several potential causes have been forwarded for amphibian population declines; the leading hypotheses have been formalized by Collins and Storfer (2003). Class I hypotheses include factors such as loss of habitat, introduction of alien species, and over-exploitation and collection of amphibians. There is accumulating evidence that stochastic variation in breeding success, coupled with isolation of certain populations by reduction in ecological corridors, may be a leading cause of extinctions at the population level (Richter et al. 2003). In the Western United States, habitat loss and fragmentation have been implicated; more than 90% of aquatic habitats have been destroyed in Southern California and Arizona. Given the rapid human population increases in California and the limited natural water sources, especially in Southern California, it is not surprising that 3 of the 10 anuran species currently listed as endangered or threatened inhabit California (USFWS 2003). In the case of the CRLF, population declines in the 20th century were preceded by the conversion of Sacramento and San Joaquin Valley wetlands and riparian habitats to agricultural land (Federal Register 2000). Streams were stripped of natural vegetation and canalized, and livestock grazing removed vegetation cover and undercut the banks along streams, causing increased water temperatures and lack of cover. These changes in riparian habitat reduced historical wetlands by more than 90%, with most habitat loss having occurred before 1939, and CRLF populations were entirely absent from the Central Valley floor by 1960. Those populations that remained in the Sierra Nevada foothills were separated from other breeding populations, further increasing the genetic isolation of remaining populations. The introduction of bullfrogs also has been linked empirically to adverse effects on survival in the CRLF. Less than 5% of CRLF tadpoles survive in ponds containing bullfrog tadpoles, whereas 30% to 40% of the tadpoles survive in the absence of bullfrog tadpoles (Lawler et al. 1999; Federal Register 2000). In a long-term study, Vredenburg (2004) demonstrated that introduction of trout into mountain ponds in the Sierras was responsible for the decline of this mountain frog (*Rana mucosa*) due to predation on tadpoles.

Class II hypotheses include global changes in climate (UV radiation, global warming), emerging diseases (such as the opportunistic chytridomycete fungus), and contaminants such as pesticides and industrial waste products. In the case of *Rheobatrachus*, there is compelling evidence for a role of the opportunistic chytridomycete fungus as a proximate cause for mass mortality in southern and northern populations of this genus (Berger et al. 1998). The temporal and geographic decline of *Rheobatrachus* and sympatric species inhabiting riparian areas of eastern Australia is consistent with an epidemic outbreak of chytrid fungus, and several recent studies confirm the association of this fungus with mass mortality events in anuran species (Berger et al. 1998; Morehouse et al. 2003). Chytrid fungus also has been associated with declines in North American populations of the boreal toad and with declines in the single remaining population of the Wyoming toad (*Bufo baxteri*) (USFWS 2001). Recently, chytrid fungus has been implicated in declining populations of the mountain yellow-legged frog (*Rana mucosa*, Fellers et al. 2001), another species inhabiting the Sierra Nevada range.

The fact that amphibians have a relatively permeable integument and generally deposit eggs with little protection from contaminants has led to speculation that they may be sensitive indicators of contaminant exposure (see Blaustein and Johnson 2003), in a way serving as "canaries in the coal mine." However, the role of contaminants in amphibian declines has been hotly debated, especially because many reports of declining species have occurred in areas that should be protected, at least in theory, from widespread agricultural or industrial contamination (USGS 2004). There are data linking airborne drift of contaminants (organophosphorus insecticides) with amphibian declines in California. Sparling et al. (2001) reported that surface waters in Sequoia National Park at an elevation (2000+ m) that had been associated with declining frog populations contained greater than 100 $ng \cdot L^{-1}$ chlorpyrifos and greater than 65 $ng \cdot L^{-1}$ diazinon. Furthermore, treefrog (*Hyla regilla*) tadpoles collected from populations in Sequoia and Yosemite National Parks, and located downwind from agricultural areas in the Sacramento and San Joaquin valleys, had body burdens of chlorinated pesticide residues 2 to 3 times greater than tadpoles from coastal areas of California (Sparling et al. 2001).

Although there have been no studies directed at examining a potential link between geographical and historical use of perchlorate and amphibian population declines, there are some data on overlap between amphibian populations and perchlorate-contaminated watersheds located in designated habitats for endangered or threatened amphibian species. As a result of cleanup efforts associated with ground water contamination in an Aerojet General Corporation's facility in Sacramento County, perchlorate was detected in ground water at concentrations as great as 8 $mg \cdot L^{-1}$. As of March 2004, perchlorate had been detected in wells and other drinking water supplies in 10 counties, with the highest concentrations reported as of March 2004 in San Bernardino and Sacramento counties, where perchlorate concentrations as high as 820 $\mu g \cdot L^{-1}$ and 400 $\mu g \cdot L^{-1}$, respectively, were reported. Perchlorate concentrations greater than 100 $\mu g \cdot L^{-1}$ have been detected in more than 139 wells in Los Angeles, Sacramento, and San Bernardino counties (California EPA 2003). Los Angeles County is listed in the proposed critical habitat designation for the CRLF (USFWS 2000). Other counties designated as containing critical habitat for the CRLF (Federal Register 2000) and for which perchlorate data are available (California EPA 2003) are Riverside (4.0 to 56 $\mu g \cdot L^{-1}$), Santa Clara (4.0 to 8.5 $\mu g \cdot L^{-1}$), San Diego (4.0 to 4.7 $\mu g \cdot L^{-1}$), and Ventura (5.7 to 20 $\mu g \cdot L^{-1}$, all measured on the U.S.N.–operated Saint Nicolas island, 34 of 34 samples containing perchlorate >4 $\mu g \cdot L^{-1}$). Thus, it appears that concentrations of perchlorate in water bodies in California measured since 1997 can be quite substantial (greater than 100 $\mu g \cdot L^{-1}$ in many cases). It is important to note, however, that the large majority of these data reflect measurements made on groundwater, and the degree to which surface waters inhabited by frogs may come in contact with contaminated ground water is an additional uncertainty.

In summary, although no studies to date have been designed to examine explicitly an association between the geographical and historical use of perchlorate and amphibian declines, data on the existing overlap between perchlorate ground water and historical species distribution can be taken into account when critical habitat for relocation and reestablishment of amphibians is being set aside.

Routes of Exposure

Perchlorate may enter surface waters from runoff or via ground water contamination and may be present in vegetation (Yu et al. 2004) eaten by larval amphibians. Another potential route of exposure is via airborne drift; sodium chlorate (ClO_3^-), which disrupts thyroid function by a mechanism similar to that of perchlorate (Hooth et al. 2001; Van Sande et al. 2003), is applied as an agricultural defoliant on cotton in West Texas (West Texas IPM update, 2002) and possibly elsewhere.

The sodium-dependent iodide symporter (NIS) is the only known mechanism for transporting perchlorate across epithelial membranes. This protein has been identified in epithelia from many extra-thyroidal tissues (see Chapter 3). Thus, the possibility exists for several possible routes of perchlorate uptake in amphibians. Theoretically, perchlorate may enter the body across epithelia in the skin or gills (in amphibian larvae) which are in direct contact with surface water or in herbivorous tadpoles via dietary intake by ingesting vegetation that has accumulated perchlorate (Yu et al. 2004) (see Figure 6-1). Preliminary studies have provided empirical evidence for perchlorate sensitive $^{125}I^-$ uptake by the gastrointestinal (GI) tract in frogs (Harrison et al. 2002; Carr et al. 2003). The elimination rate of perchlorate in frogs is not known, although based on studies in mammals, perchlorate should be eliminated relatively rapidly. In *Xenopus laevis* larvae, the effects of a 70-d perchlorate exposure are reversed within 28 d of non-treatment (Goleman et al. 2002b). A conceptual model illustrating environmental effects and fate of perchlorate in amphibians is in Figure 6-2.

Amphibians must obtain and sequester iodide from their environment in order to synthesize thyroid hormones (THs). Because iodide and perchlorate are transported by the same protein, it is logical that the degree to which either anion is taken up will depend upon the relative abundance of the other (Figure 6-1B). Simply stated, the effect of perchlorate will depend upon the availability of iodide to the organism. There is empirical evidence to support this hypothesis. Sparling et al. (2003) found that simultaneous exposure to iodide reduced the antimetamorphic effects of

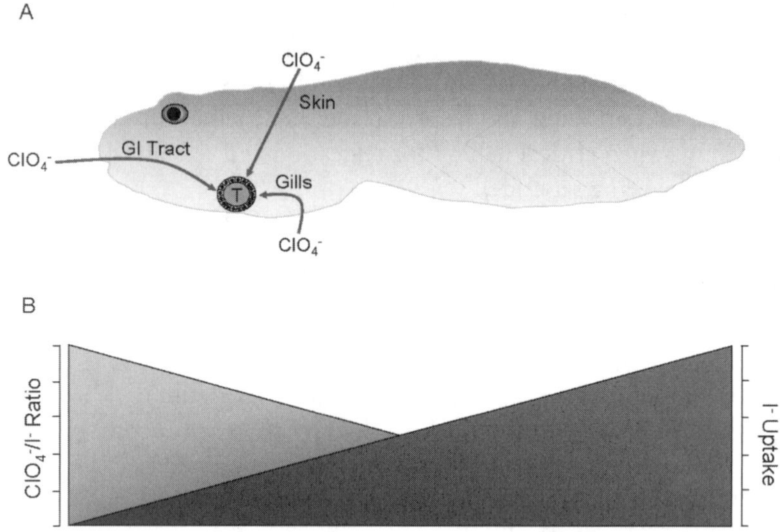

Figure 6-1 A) Potential sites for perchlorate uptake in larval frogs. Perchlorate may be taken up by a sodium-dependent iodide symporter (NIS) in a number of epithelia including skin, GI tract, and respiratory epithelium. Once in the bloodstream, perchlorate travels to the thyroid gland (T) to block iodide uptake (also see Chapter 3). B) The efficiency of iodide uptake by a larval frog will depend upon the relative amounts of perchlorate and iodide in the environment. As perchlorate in the environment increases, the ability to take up iodide decreases proportionately.

perchlorate exposure in tree frogs. Hu et al. (2003) found that increasing iodide availability eliminated the inhibitory effects of perchlorate on forelimb emergence, tail absorption, and hindlimb growth in *X. laevis* (Table 6-1).

Hypothalamus–Pituitary–Thyroid Axis in Amphibians

In amphibians, there are paired thyroid glands attached to the hyoid cartilage, and TH synthesis is thought to occur much as it does in mammals. For example, organification of iodide is blocked by propylthiouracil (PTU) (Burd 1992) and methimazole (Becker et al. 1997) in amphibians just as in mammals. In addition, thyroglobulin (Tg) structure appears to have been phylogenetically conserved in those amphibians in which the protein has been purified or the gene for Tg cloned. Amphibian and reptilian THs are structurally identical to those in mammals and in all other vertebrate classes, and techniques (such as radioimmunoassay, high-performance liquid chromatography) for measuring THs in mammalian

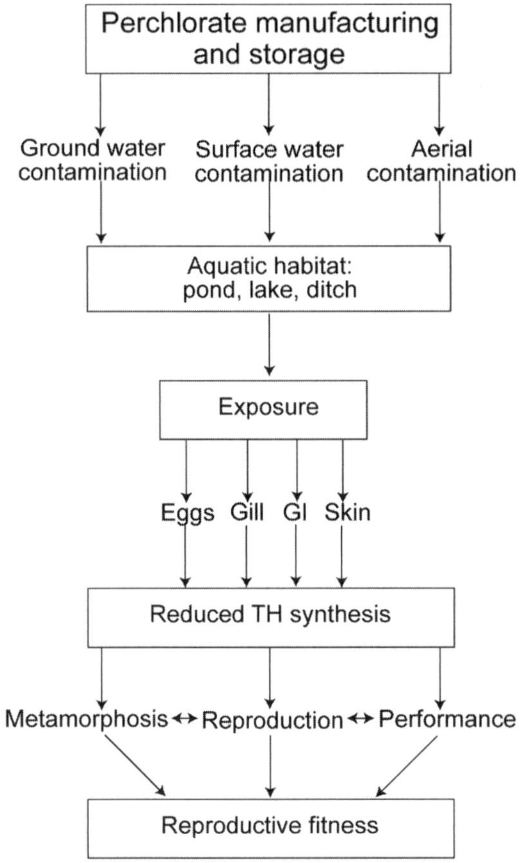

Figure 6-2 Conceptual model illustrating potential sources of perchlorate and potential outcomes of perchlorate exposure in amphibians

species can be applied successfully to amphibians and reptiles. Taken together, these findings indicate that pathways for TH synthesis in amphibians and reptiles are similar if not identical to the pathways in mammals.

It is generally believed that THs are transported in the blood attached to transport proteins in amphibians and reptiles, similar to the situation in mammals. Transthyretin mRNAs and proteins have been sequenced or purified in amphibian (Prapunpoj et al. 2000; Yamauchi et al. 2002) and reptilian (Prapunpoj et al. 2002) species, suggesting that a transthyretin-like binding protein has been conserved evolutionarily. Both T_3 and T_4 are present in the circulation of amphibians and reptiles, and deiodination is presumed to be an important means of hormone activation and deactivation, as in mammals. Homologs of the mammalian 5'DI, 5'DII, and 5D are present in amphibians (Becker et al. 1997).

The function of the hypothalamo–hypophysial–thyroid axis has been essentially conserved in amphibians in the sense that thyroid gland function is controlled primarily by thyroid-stimulating hormone (TSH; Dodd and Dodd 1976). mRNAs encoding both the alpha and the beta subunits of TSH have been sequenced in *X. laevis* (Buckbinder and Brown 1993). Recently, Okada et al. (2004) developed a homologous radioimmunoassay (RIA) for TSH in bullfrogs, facilitating the ability to accurately measure plasma TSH in anurans. Given the abundant evidence that perchlorate

Table 6-1 Mean incidence of forelimb emergence, complete tail absorption, and hindlimb length in *Xenopus laevis* tadpoles exposed to FETAX[a] medium or ammonium perchlorate in the presence or absence of sodium iodide for 70 d

Treatment	Forelimb emergence (FLE)[b] (% ± SEM)	Complete tail absorption[c] (% ± SEM)	Hindlimb length (mm)
Frog embryo teratogenesis assay (FETAX) control	91.6 ± 1.83	91.1 ± 2.44	19.1 ± 0.19 (n = 158)
14 mg·L^{-1} ammonium perchlorate (AP)	0.00 ± 0.00[d]	0.00 ± 0.00[d]	2.45 ± 0.07[d] (n = 167)
14 mg·L^{-1} sodium iodide (NaI)	88.3 ± 3.61	84.5 ± 5.68	20.1 ± 0.22 (n = 147)
14 mg·L^{-1} NaI + 14 mg·L^{-1} AP	94.3 ± 1.21	92.1 ± 0.66	19.0 ± 0.19 (n = 162)

[a] Controls were reared in untreated FETAX (*Xenopus*, Dawson and Bantle 1987). Exposures began within 24 h of fertilization and continued for 70 d. Sample size of 4 tank replicates for incidence data. Sample size for hindlimb length measurements is indicated in parentheses. Mortality during the 70-d exposure was <6% for all groups except the NaI treatment, in which mortality was 12%.

[b] Forelimb emergence, number with forelimbs divided by number hatching during the 70-d exposure. Hatching exceeded 87% in all groups.

[c] Number with complete tail absorption divided by number hatchings.

[d] Significantly different from FETAX medium control ($p < 0.05$).

elevates blood TSH levels in other organisms (due to removal of feedback inhibition by thyroxine), measurement of blood TSH in amphibians will be an important tool for assessing perchlorate exposure in future studies.

In adult amphibians, thyrotropin-releasing hormone (TRH) acts on the anterior pituitary gland to stimulate TSH secretion (Denver 1988; Denver and Licht 1989). Hypothalamic corticotropin-releasing factor (CRF) also appears to be an important stimulator of pituitary TSH secretion in amphibians. Treatment of frog and hatchling turtle pituitaries with CRF elicits TSH secretion (Denver and Licht 1989; Boorse and Denver 2004; Okada et al. 2004). In larval amphibians, TSH secretion is not stimulated by TRH (Denver and Licht 1989), and CRF is the principal hypothalamic regulator of the pituitary–thyroid axis during metamorphosis (Denver et al. 2002).

Role of Thyroid Hormones in Reproduction and Development in Amphibians

Thyroid hormones and amphibian metamorphosis

A unique and dramatic feature of normal life history for many amphibian species is a postembryonic development phase called "metamorphosis," in which the organism transforms from an aquatic larval form to a terrestrial or semiterrestrial form. This requires normal TH production, which is why many studies have focused on the impact of perchlorate exposure on metamorphosis. Thyroid hormones are required for the massive reprogramming of gene expression that is required for the morphological, biochemical, and physiological changes that take place during metamorphosis, including resorption of the larval tail, reorganization of the GI tract, development of limbs, restructuring of the skeleton, and a switch in nitrogen metabolism from ammonia to urea production (Shi 2000).

A consequence of perchlorate disruption of TH synthesis is that less TH is available for interaction with TH receptors. Amphibians possess TH receptors that are similar in structure and function to their mammalian counterparts, and receptor-mediated gene transcription is the major form of TH action during metamorphosis. It is generally believed that THs interact with their receptors to control metamorphosis by directly influencing the expression of immediate early-response genes whose products, in turn, affect the expression of late-response genes, resulting in a gene regulation cascade (Shi 2000). Many of the early TH response genes encode transcription factors (including the TRß gene) that, in turn, repress or activate the expression of late response genes (Shi 2000) during metamorphosis.

There are many hormones that can act to modify TH action during metamorphosis. Corticosteroids (corticosterone, aldosterone) secreted from the larval interrenal tissue have been reported to accelerate TH-induced metamorphosis, although the effects are stage and tissue dependent. Corticosteroid levels in blood are elevated during metamorphosis, and receptors for these steroids are present in a number of larval organs, including tail and liver (see Krain and Denver 2004 for a recent review). Prolactin (PRL), a pituitary hormone, has been reported to inhibit TH-induced metamorphosis. A physiological role for PRL in metamorphosis remains a matter of debate, however, because the production of PRL and its receptor are elevated during normal metamorphosis (Buckbinder and Brown 1993).

Not all amphibians must complete metamorphosis in order to reproduce. For example, some urodele species reach reproductive maturity and breed in a larval form, a condition generally known as "neoteny." Some species (the tiger salamander, *Ambystoma tigrinum* and the Mexican axolotl, *Ambystoma mexicanum*, e.g.) undergo metamorphosis only under favorable environmental conditions. This phenomenon, called "facultative neoteny," cannot be explained by a lack of functional TH receptors because the axolotl has functional alpha and beta forms of the receptor (Safi et al. 2004). Other species never undergo metamorphosis, a condition referred to as "obligate neoteny." Unlike facultative neotenes, obligate neotenes (such as the mudpuppy *Necturus maculosus*) do not respond to TH treatment by initiating metamorphosis. Neoteny has not been reported to date for any anuran species.

Thyroid hormone production and secretion increases during metamorphosis, coinciding with dramatic morphological changes such as emergence of the forelimbs and tail resorption. Expression and activity of a 5' deiodinase (Type II or D2) increases during metamorphosis (Becker et al. 1997; Manzon and Denver 2004) in a tissue-specific manner, presumably resulting in more circulating T_3, although some there are some data to indicate tissue specific increases during metamorphic climax in the activity of type III deiodinase (Manzon and Denver 2004), which inactivates T_4 and T_3 (see Chapter 3). Early attempts to induce TH secretion and metamorphosis with TRH were unsuccessful, suggesting that control of the pituitary–thyroid axis in larval amphibians was very different than what was known to occur in mammals. There is now substantial evidence that CRF exerts dual control over the pituitary–thyroid and pituitary–adrenal axes during metamorphosis. Administration of CRF to tadpoles elevates plasma thyroxine and corticosterone levels (Gancedo et al. 1992; Denver 1993; Boorse and Denver 2004) and accelerates metamorphosis in a number of anuran and urodeles species (Miranda et al. 2000; Boorse and Denver 2002, 2004). Levels of immunoreactive CRF in the median eminence increase during the latter stages of metamorphosis (Carr and Norris 1990), further implicating this peptide in the hypothalamic regulation of thyroid function.

As described later in this chapter, disruption of thyroid function by perchlorate or other contaminants can affect the rate of metamorphosis or limb growth. There are important ecological consequences to interrupting or delaying metamorphosis. The length of the larval period can vary tremendously within amphibians, taking up to 3 years in some species such as the American bullfrog (*Rana catesbeiana*) or lasting just a few weeks in some desert-adapted species inhabiting ephemeral ponds, such as in species of spadefoot toads that inhabit the southwestern United States (genera *Scaphiopus* and *Spea*, family Pelobatidae). Delaying or disrupting the timing of metamorphosis can have potentially dramatic effects on organism performance and may ultimately affect fitness by making larvae more susceptible to aquatic predators. In desert-dwelling species, a delay in the timing of metamorphosis may decrease the likelihood that the animals will complete metamorphosis before the pond dries up. The potential ecological consequences of delaying metamorphosis are illustrated in Figure 6-3.

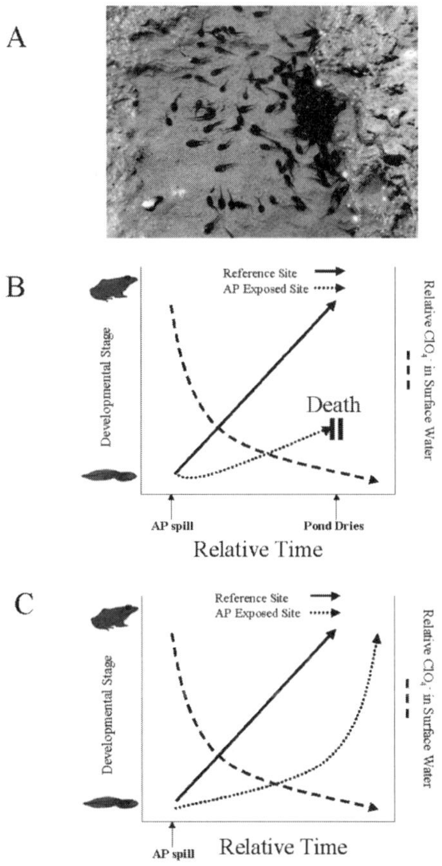

Figure 6-3 Perchlorate inhibition of metamorphosis may have different consequences for tadpoles with short and long larval periods. A) Tadpoles inhabiting ephemeral ponds must develop rapidly before the pond dries up. B) Perchlorate effects on metamorphosis in a species with a short larval period. Perchlorate inhibition of metamorphosis may prevent the animals from leaving the pond before it dries up. C) Perchlorate effects on metamorphosis in a species with a long larval period. Perchlorate exposure earlier in development may delay, but not prevent, metamorphosis because the animals would recover from perchlorate exposure over time.

Thyroid hormones and reproduction in amphibians

Normal TH secretion is required for masculinization in anurans based upon data collected primarily in *X. laevis*. Thyroid hormones appear to be necessary for expression of androgen receptors in *X. laevis* (Cohen and Kelley 1996; Robertson and Kelley 1996), and blocking normal TH synthesis can cause feminization of offspring (Hayes 1997; Goleman et al. 2002b). The link between THs and reproductive development is less clear for urodeles because neotenic salamanders breed successfully year after year in some locales. No clear link has been established between seasonal changes in thyroid activity and sex steroid levels or gonadal function in anurans (Vandorpe et al. 1990), although some (Tasaki et al. 1986; Gancedo et al. 1995) have reported sex differences in seasonal cycles of plasma T_3 and T_4 in frogs.

Other roles for thyroid hormones in amphibians

In mammals and birds, THs increase oxygen consumption in virtually every organ in the body by influencing cellular respiration via a number of long- and short-term actions on mitochondrial energy transfer. Amphibians do not produce significant amounts of heat, and data suggesting a link between THs and oxygen consumption in amphibians are equivocal, with some studies reporting an effect (May et al. 1976; Gupta and Deka-Borah 1995) and others reporting no effect (Gupta and Chakrabarty 1990) of THs on respiration in amphibians.

Thyroid hormones play an important role in brain development in amphibians as they do in mammals. Thyroid hormones are required for proper development of the median eminence and hypothalamus, and specifically play a role in the development of catecholaminergic cells groups in the amphibian brain (Kikuyama et al. 1979; Carr et al. 1991; Norris et al. 1992), very similar to the role they play in the development of catecholaminergic pathways in the mammalian brain. Thyroid hormones also are required for development of the cerebellum (Hauser and Gona 1983, 1984; Hauser et al. 1986) and visual pathways in frogs (Hoskins and Grobstein 1985).

Perchlorate Effects on Amphibians

Mechanism of perchlorate action in amphibians

Although a full-length NIS has yet to be cloned from any amphibian or reptilian species, indirect evidence indicates that perchlorate acts to blocks iodide transport across the basaolateral plasma membrane of thyroid follicle cells just as has been shown in mammals. Partial cDNA fragments encoding putative NIS proteins have been sequenced from whole *X. laevis* tadpoles (Carr et al. 2003) and isolated bullfrog tadpole thyroid glands (unpublished results). Administration of excess iodide can prevent the inhibition of perchlorate-induced inhibition of TH synthesis in amphibians, presumably by competing with perchlorate for the NIS on thyroid follicle cells (Hu et al. 2003; Sparling et al. 2003).

Acute toxicity of perchlorate in amphibians

Perchlorate is not overtly toxic to adult or larval frogs at concentrations that are likely to be encountered in the environment. Studies on the lethality of ammonium perchlorate (AP) in embryonic and larval *X. laevis* have reported LC50s ranging from 496 $mg \cdot L^{-1}$ to 510 $mg \cdot L^{-1}$ (Bantle et al. 1999; Goleman et al. 2002a). In a more recent study, Dean et al. (2004) examined the acute toxicity of sodium perchlorate in premetamorphic green frog tadpoles. The LC50 for 96 h mortality for sodium perchlorate in green frog tadpoles was 5500 $mg \cdot L^{-1}$ (Dean et al. 2004).

Perchlorate effects on amphibian metamorphosis

The ability of perchlorate salts to inhibit metamorphosis has been known and exploited as a research tool for decades. Given the known mode of action for perchlorate, it is not surprising that perchlorate exposure affects thyroid follicle cell structure. Miranda et al. (1996) reported that exposure of *Bufo arenarum* larvae to perchlorate caused hypertrophy of thyroid follicle cells and an increase in thyroid gland volume. Consistent with the reported effects of perchlorate on thyroid follicle cell structure, high concentrations of perchlorate salts block metamorphosis and prevent normal development of the immune system in *X. laevis* (Rollins-Smith et al. 1990, 1993, 1997). Interestingly, perchlorate also interferes with the development of hardwired connections between the central nervous system and the immune system by disrupting the normal development of catecholaminergic innervation to immune organs such as the spleen (Kinney et al. 1996). The effects of perchlorate on development of the neuroendocrine system are complex, as some processes such as development of thyrotropes and lactotropes are altered whereas the development of other systems, such as the CRF pathway, does not appear to be affected (Miranda and Dezi 1997). However, this latter finding (or lack of response) has not yet been confirmed by other laboratories and is not consistent with the observed effects of thyroid inhibition on the development of hypophysiotropic pathways. For example, Shi et al. (1994) reported that normal, circulating THs are required for development of the

CRF and pro-opiomelanocortin neuronal systems in rats. Given the well-known role of THs in controlling the development of the central nervous system (CNS), more work needs to be done on the effects of perchlorate on CNS development in amphibians.

The majority of studies conducted on perchlorate and amphibian metamorphosis have employed large concentrations of perchlorate as a tool for inhibiting metamorphosis, but have not addressed the exposures were environmentally relevant. This is understandable, as it wasn't until the late 1990s that the extent to which water supplies were contaminated with perchlorate became known (see Chapter 1). Some early studies reported that perchlorate could be detected in drinking water supplies across the country (Urbansky 2002). These findings spawned a more intensive look at perchlorate in surface and ground waters associated with military and aerospace facilities. Subsequent studies revealed that surface waters that serve as breeding ponds for amphibians may have relatively high perchlorate concentrations, in some cases as great as 30 $mg \cdot L^{-1}$ (Smith et al. 2001; USACE 2004). These studies prompted two new questions. First, do environmentally relevant concentrations of perchlorate interfere with metamorphosis? Second, can disruption of thyroid function in amphibians serve as a diagnostic tool predicting exposure consequences in other wildlife and humans? Recent research by independent laboratories indicates that environmentally relevant concentrations of perchlorate alter thyroid function, metamorphosis, limb growth and reproductive development in the laboratory. Exposure to AP in the 5 to 100 $\mu g \cdot L^{-1}$ range for 70 d can delay metamorphosis, while concentrations greater than 147 $\mu g \cdot L^{-1}$ completely inhibit metamorphosis (Goleman et al. 2002a). Using a 14-d exposure (compared to the 70-d exposure used by Goleman et al. 2002a, 2002b), Tietge et al. (2005) demonstrated that exposure to 250 μg perchlorate$\cdot L^{-1}$ inhibited metamorphosis in *X. laevis*. Perchlorate has a potent inhibitory effect on hindlimb growth within a wide range of environmental concentrations (Goleman et al. 2002a, 2002b) but does not influence overall organismal growth at concentrations up to 14 $mg \cdot L^{-1}$ (Goleman et al. 2002b). Because of the differential effects of perchlorate on hindlimb growth and somatic growth, the ratio of hindlimb length to snout-vent length has been used as a non-invasive measure of perchlorate exposure in larval frogs (Carr et al. 2003).

Among the most sensitive measures of perchlorate exposure in developing frogs are changes in thyroid histology, specifically follicle cell height and colloid depletion, both of which occur not as a direct result of perchlorate action but because of secondary or indirect effects on TSH secretion due to interference with normal negative feedback mechanisms. Exposure of developing frogs to AP results in colloid depletion and follicle cell hypertrophy (Goleman et al. 2002b; Carr et al. 2003) (Figure 6-4). Exposure of larval *X. laevis* to sodium perchlorate as low as 16 μg·L^{-1} beginning at stage 51 or 54 for just 8 d is sufficient to cause colloid depletion, while exposure for 14 d to 16 μg perchlorate·L^{-1} results in follicle cell hypertrophy as well as colloid depletion (Tietge et al. 2005). The effects of perchlorate on thyroid follicle cell hypertrophy and colloid depletion are linearly related within the concentration range of 16 to 4000 μg·L^{-1} after either 8- or 14-d exposure beginning at stages 51 or 54 (Tietge et al. 2005). Collectively, these findings show that environmentally relevant concentrations of perchlorate can inhibit metamorphosis in the surrogate species *X. laevis*. Furthermore, these findings demonstrate that thyroid histopathological indicators of perchlorate exposure are extremely sensitive, with responses occurring at the lowest perchlorate concentrations that have been tested to date. Whether concentrations less than 16 μg perchlorate·L^{-1} also produced thyroid changes remains to be tested.

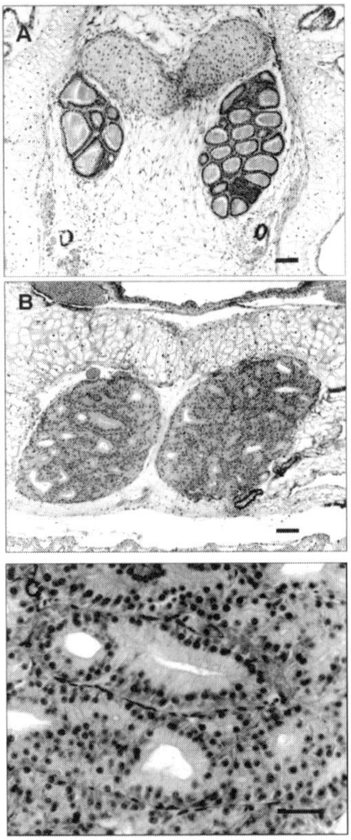

Figure 6-4 Changes in thyroid histology in a larval tadpole exposed to a high (14 mg·L^{-1}) but environmentally relevant concentration of ammonium perchlorate. A) Appearance of thyroid gland in a control animal. Calibration bar = 100 μm. B) Appearance of thyroid gland in a perchlorate-exposed tadpole. Calibration bar = 100 μm. C) Higher-magnification photomicrograph of tissue in B. Calibration bar = 50 μm. Note pronounced colloid depletion and hypertrophy of thyroid follicle cells. Similar changes have been observed in wild-caught tadpoles exposed to perchlorate under natural conditions (Carr et al. 2003).

As with other endocrine-disrupting chemicals, the response of frogs to perchlorate depends upon timing, both with respect to exposure duration and the development stage during exposure. As mentioned previously, the antimetamorphic effects of perchlorate are reversible in *X. laevis*, suggesting that amphibians may recover from perchlorate exposure, although certain developmental processes such as hindlimb growth and completion of metamorphosis may be delayed. In *X. laevis*, perchlorate exposure late in metamorphosis is not as effective at inhibiting metamorphosis or hindlimb growth exposure early in development (Hu et al. 2003; Table 6-2). Furthermore, the Tier I Frog Metamorphosis Assay for thyroid disruption as originally proposed by the Endocrine Disruptor Screening and Testing Advisory Committee (EDSTAC; Federal Register 1998) is relatively insensitive for detecting perchlorate effects because the assay test exposure occurs in a very late stage (Nieuwkoop-Faber stage 60) *X. laevis* (Goleman et al. 2002a). It is presumed that late-stage tadpoles are relatively insensitive to perchlorate and other contaminants that act directly upon the thyroid gland because blood levels of T_4 are already elevated by stage 60 and genetic reprogramming of T_4-sensitive tissues has already begun by the time forelimb emergence begins (Shi 2000). Other evidence indicates that adult frogs may not be as sensitive as larval frogs to perchlorate. Exposure of adult female *X. laevis* for 10 weeks to perchlorate concentrations as high as 14 $mg \cdot L^{-1}$ failed to cause thyroid follicle cell hypertrophy or colloid depletion (Goleman et al. 2004). Exposure of larval *X. laevis* to these conditions produces profound alterations in thyroid histopathology (see Figure 6-4).

Perchlorate effects on reproduction in amphibians

Thyroid hormones are required for masculinization in some anurans. Specifically, THs appear to be important for gonadal and secondary sex determination, possibly by inducing androgen receptor expression in androgen sensitive tissues. Exposure to perchlorate (Goleman et al. 2002a) or other goitrogenic chemicals such as PTU (Hayes 1997) can cause feminization of male offspring, producing ovaries in genetically male individuals.

Oxidative stress and susceptibility to UV radiation

The interaction between chemical exposure and UV has been the subject of recent study, and one of the proposed mechanisms of the decline of amphibians and increased incidence of deformities is an interaction between UV and chemical exposures (Schmidt 1997; Ankley et al. 1998). Thus, an increase in UV exposure is not necessarily a prerequisite of increased incidence of malformations or a decrease in population levels because it could be that an increased exposure to chemicals makes the organisms more susceptible to the effects of UV by interfering with their ability to cope with solar radiation. In fact, an in situ study of groundwater toxicity in amphibians (Bruner et al. 1998) found that the acute toxicity and magnitude of developmental effects depended on the amount of ambient solar radiation (i.e., the water was more toxic on sunny days than on cloudy days). Thus, ecological risk assessments that rely solely on

Table 6-2 Mean hindlimb length in larval *X. laevis* exposed to ammonium perchlorate beginning at different developmental stages and continuing for 68-d post-hatch

	Hindlimb length (mm ± SEM)	
Nieuwkoop-Faber stage[a]	Control[b]	Treatment (14 $mg \cdot L^{-1}$ AP)
NF 1-10	19.2 ± 0.50	3.60 ± 0.33
NF 49	13.2 ± 0.51	4.97 ± 0.38
NF 55	17.9 ± 0.29	13.8 ± 0.48

[a] Stage at which exposure began (Nieuwkoop and Faber 1994). Exposures continued for 68-d post-hatch.

[b] Controls were reared in untreated Frog Embryo Teratogenesis Assay (FETAX)–*Xenopus* (Dawson and Bantle 1987)

measurement of environmental concentrations of chemicals and exposure of laboratory organisms to these concentrations may be biased if it does not take into account interactions between chemicals and natural stressors such as UV. For example, it has been suggested that, because development of the amphibian skin and melanocytes is mediated by THs, and because one function of melanocytes is to protect the animals against UV exposure, then exposure of organisms to thyroid-disrupting chemicals may increase an amphibian's susceptibility to UV (Burkhart et al. 1999). Indeed, recent evidence has found that exposure of *Xenopus* larvae to perchlorate dramatically enhanced developmental deformities in UV-exposed individuals (Burkhart et al. 2000).

In order to test this hypothesis, *X. laevis* tadpoles were exposed to 2 doses of sodium perchlorate and UV radiation. The doses of perchlorate used were 0.05 and 10 $mg \cdot L^{-1}$, concentrations that were found in previous studies to inhibit and arrest metamorphosis in this species, respectively (Goleman et al. 2002b). The doses of UV were similar to those found in natural sunlight. The endpoints were levels of DNA strand breakage, oxidative DNA damage, and UV-induced photodimers. It was found that both perchlorate and UV exposure induced oxidative damage in the DNA, and that this may have been influenced by the levels of ammonia that accumulated as waste products in the exposure beakers. In addition, mortality was strongly correlated with the amount of oxidative DNA damage present. Furthermore, the amount of UV dimers in the DNA was enhanced by treatment with perchlorate, at least in some treatments (McDaniel 2004).

Field studies

An initial assessment of perchlorate in water, soil, and animal tissue samples at Longhorn Army Ammunition Plant (LHAAP) in East Texas revealed that perchlorate concentrations in tissues from frogs and tadpoles ranged from nondetectable to >500 μg perchlorate$\cdot$kg^{-1} in animals collected from 1 pond where water concentrations of the contaminant were as great as 30 mg$\cdot$L^{-1} (Smith et al. 2002). American bullfrog tadpoles collected from this site had significantly shorter hindlimbs but identical body length compared to tadpoles collected from a reference site with no detectable perchlorate contamination (Carr et al. 2003). In anuran tadpoles, hindlimb growth requires normal TH secretion, and Goleman et al. (2002a) demonstrated that perchlorate exposure causes a concentration-dependent reduction in hindlimb growth in *X. laevis*, presumably due to reduced TH synthesis (Goleman et al. 2002b). Animals from the contaminated site at LHAAP were at an earlier stage in metamorphosis than animals from the control site (Carr et al. 2003). Histological analysis of thyroid glands from tadpoles collected from the contaminated site, however, revealed smaller thyroid glands, a finding that is not consistent with the increase in thyroid gland size that one would expect to occur after perchlorate exposure. However, the comparison may have been confounded by the fact that the thyroid gland increases in size normally during metamorphosis, and as a result, animals from the reference site may have had larger thyroid glands because they were, on average, at more advanced stages of metamorphosis (Carr et al. 2003). To avoid a confounding effect of developmental stage on the analysis of thyroid tissue response to perchlorate, Carr et al. (2003) examined thyroid histology in stage-matched chorus frog tadpoles inhabiting reference and contaminated sites at LHAAP. Comparison of stage-matched chorus frog tadpoles from a perchlorate and contaminated reference site indicated gross histological changes in the thyroid gland that were consistent with perchlorate exposure. Tadpoles from the contaminated site had significantly greater colloid depletion and greater hypertrophy of thyroid follicle cells compared to tadpoles from the reference site (Carr et al. 2003). The follicle cell hypertrophy and colloid depletion observed in the tadpoles from the contaminated site are consistent with histopathological changes reported by others (Goleman et al. 2002b; Tietge et al. 2005).

There is only one report of potential thyroid disruption in response to perchlorate exposure under field conditions. Theodorakis et al. 2006 reported a correlation between perchlorate concentrations and thyroid follicle cell height, but not colloid depletion, in adult male cricket frogs (*Acris crepitans*) collected from the Lake Waco and Lake Belton watersheds. Cricket frogs inhabiting streams contaminated with 10 to 30 μg perchlorate·L^{-1} exhibited follicle cell hypertrophy when compared to frogs inhabiting streams with 5 μg perchlorate·L^{-1} or less (Theodorakis et al. 2006).

Research Recommendations

In order to understand the mechanism of perchlorate action in amphibians, the amphibian NIS must be characterized in more detail, both with respect to molecular structure and the 1st-order rate kinetics for iodide and perchlorate transport by this protein. Establishment of kinetic parameters will allow researchers to determine how exposure endpoints such as blood levels of perchlorate might interfere with iodide transport. Moreover, the role of the NIS in extrathyroidal tissues needs to be better understood in order to characterize the most likely routes of perchlorate exposure in amphibians. Furthermore, the role that environmental iodide plays in determining the response of amphibians to perchlorate needs to be better elucidated. At the very least, the relative availability of iodide to the organism must be considered when perchlorate effects are examined. Although our research indicates that perchlorate may interfere with normal gonadal differentiation (Goleman et al. 2002b), the implications of this result for reproductive success and population stability are unknown. Although population modeling holds promise for testing the population-level effect of perchlorate, multigenerational exposure studies should also be conducted to gather empirical data on the biological significance of perchlorate-induced changes in sex ratio. Finally, recent data indicate effects of perchlorate on thyroid histology at concentrations as low as 16 μg·L^{-1} (Tietge et al. 2005). However, as with other studies (Goleman et al. 2002b), histological changes in the thyroid occur at the lowest concentrations of perchlorate tested, and a true lowest-observable-effect concentration (LOEC) has yet to be determined, although it appears to be less than 16 μg perchlorate·L^{-1}. Future studies should include perchlorate concentrations sufficiently low to bracket an LOEC for perchlorate on thyroid histology in tadpoles.

Conclusions

Amphibian metamorphosis is a unique model for examining the role of THs in development and growth. Amphibians require TH for normal growth, development, and, at least in anurans, normal reproductive development. At present, there are no empirical data linking perchlorate exposure to amphibian declines. However, given the large database on perchlorate levels in surface waters, potential exposure can and may be considered when determining critical habitat designations for threatened or endangered amphibian species. Environmentally relevant concentrations of perchlorate alter the rate of metamorphosis in a reversible fashion. The antimetamorphic effects of perchlorate may have important ecological effects, especially in those species inhabiting ephemeral ponds and having short developmental rates. Many of the same endpoints used to gauge perchlorate exposure in other vertebrates have been used successful to assess perchlorate exposure in amphibians, with the most sensitive being histopathological endpoints. These endpoints have been used successfully to assess perchlorate exposure under both laboratory and field conditions. Recent data also indicate that the response to perchlorate depends upon the timing of exposure, both in the length of exposure and the period during development when exposure occurs. Adult frogs are not as sensitive to perchlorate exposure as are developing frogs, and tadpoles at later stages of development are not as sensitive to the antimetamorphic effects of perchlorate as are tadpoles at earlier stages of development. Amphibians represent a unique and powerful model for examining the organismal response to perchlorate exposure and the potential effects of this contaminant on thyroid homeostasis, reproduction, and epithelial iodide transport.

References

Ankley GT, Tietge JE, DeFoe DL, Jensen KM, Holcombe GW, Durhan EJ, Diamond SA. 1998. Effects of ultraviolet light and methoprene on survival and development of *Rana pipiens*. Environ Toxicol Chem 17:2530–2542.

Bantle JA, Dumont JN, Harvey JG, Mattie DR. 1999. FETAX assay of ammonium perchlorate. Toxicologist 48:528.

Becker KB, Stephens KC, Davey JC, Schneider MJ, Galton VA. 1997. The type 2 and type 3 iodothyronine deiodinases play important roles in coordinating development in *Rana catesbeiana* tadpoles. Endocrinology 138:2989–2997.

Berger L, Speare R, Daszak P, Green DE, Cunningham AA, Goggin CL, Slocombe R, Ragan MA, Hyatt AD, McDonald KR, Hines HB, Lips KR, Marantelli G, Parkes H. 1998. Chytridiomycosis causes amphibian mortality associated with population declines in the rain forests of Australia and Central America. Proc Natl Acad Sci 95:9031–9036.

Blaustein AR, Johnson PTJ. 2003. The complexity of deformed amphibians. Front Ecol Environ 1:87–94.

Boorse GC, Denver RJ. 2002. Acceleration of *Ambystoma tigrinum* metamorphosis by corticotropin-releasing hormone. J Exp Zool 293: 94–98.

Boorse GC, Denver RJ. 2004. Expression and hypophysiotropic actions of corticotropin-releasing factor in *Xenopus laevis*. Gen Comp Endocrinol 137:272–282.

Bruner MA, Rao M, Dumont JN, Hull M, Jones T, Bantle JA. 1998. Ground and surface water developmental toxicity at a municipal landfill: Description and weather-related variation. Ecotoxicol Environ Saf 39:215–226.

Buckbinder L, Brown DD. 1993. Expression of the *Xenopus laevis* prolactin and thyrotropin genes during metamorphosis. Proc Natl Acad Sci 90:3820–3824.

Burd GD. 1992. Development of the olfactory nerve in the clawed frog, *Xenopus laevis*: II. Effects of hypothyroidism. J Comp Neurol 315:255–263.

Burkhart JG, Cline JE, Harmon G, Fort DJ. 2000. Possible involvement of thyroid and retinoid receptor disruption in amphibian malformation. Society of Environmental Toxicology and Chemistry 21st Annual Meeting; 2001 Nov 11–16; Nashville, TN.

Burkhart JG, Cline JE. 1999. Interaction among environmental factors that contribute to malformation in frogs. Society of Environmental Toxicology and Chemistry 20th Annual Meeting; 1999 Nov 14–18; Philadelphia, PA.

[California EPA] California Environmental Protection Agency, Department of Health Services. March 2004. Perchlorate database. Available from http://www.dhs.ca.gov/ps/ddwem/chemicals/perchl/perchlindex.htm. Accessed 6 Apr 2006.

Carr DL, Laharrague F, Kahn B, Pressley TA, Carr JA. 2003. Molecular characterization of a putative sodium/iodide symporter in the South African clawed frog, *Xenopus laevis*. Ann NY Acad Sci 986:711–712.

Carr JA, Norris DO. 1990. Immunohistochemical localization of corticotropin releasing factor and arginine vasotocin like immunoreactivities in the brain and pituitary of the American Bullfrog (*Rana catesbeiana*) during development and metamorphosis. Gen Comp Endocrinol 78:180–188.

Carr JA, Norris DO, Samora A. 1991. Organization of tyrosine hydroxylase-immunoreactive neurons in the di- and mesencephalon of the American bullfrog (*Rana catesbeiana*) during metamorphosis. Cell Tissue Res 263:155–163.

Carr JA, Urquidi LJ, Goleman WL, Hu F, Smith PN, Theodorakis CW. 2003. Ammonium perchlorate disruption of thyroid function in natural amphibian populations: Assessment and potential impact, In Linder G, ed, Multiple Stressor Effects in Relation to Declining Amphibian Populations, ASTM STP 1443, ASTM International, West Conshohocken, PA, USA, pp 131–142.

Cohen MA, Kelley DB. 1996. Androgen-induced proliferation in the developing larynx of *Xenopus laevis* is regulated by thyroid hormone. Dev Biol 178:113–123.

Collins JP, Storfer A. 2003. Global amphibian declines: sorting the hypotheses. Diversity Distributions 9: 89–98.

Dawson DA, Bantle JA. 1987. Development of a reconstituted water medium and preliminary validation of the frog embryo teratogenesis assay—Xenopus (FETAX). J Appl Toxicol 7:237–244.

Dean KE, Palacheck RM, Noel JM, Warbritton R, Aufderheide J, Wireman J. 2004. Development of freshwater water-quality criteria for perchlorate. Environ Toxicol Chem 23:1441–1451.

Denver RJ. 1988. Several hypothalamic peptides stimulate in vitro thyrotropin secretion by pituitaries of anuran amphibians. Gen Comp Endocrinol 72:383–393.

Denver RJ. 1993. Acceleration of anuran amphibian metamorphosis by corticotropin-releasing hormone-like peptides. Gen Comp Endocrinol 91:38–51.

Denver RJ, Boorse GC, Glennemeier KA. 2002. Endocrinology of complex life cycles: amphibians. In: Pfaff D, Arnold A, Etgen A, Fahrbach S, Moss R, Rubin R, editors. Hormones, brain and behavior. San Diego (CA): Academic. Volume 2, p 469–513.

Denver RJ, Licht P. 1989. Neuropeptide stimulation of thyrotropin secretion in the larval bullfrog: evidence for a common neuroregulator of thyroid and interrenal activity during metamorphosis. J Exp Zool 252:101–104.

Dodd MHI, Dodd JM. 1976. The biology of metamorphosis. In: Moore JA, editor. Physiology of the amphibia. New York (NY): Academic. Volume 3, p 467–599.

Feder ME. 1992. A perspective on environmental physiology of the amphibians. In: Feder ME, Burggren WW, editors. Environmental physiology of the amphibians. Chicago (IL): Univ Chicago Pr. p 1–6.

Federal Register. 1998. Endocrine disruptor screening program:Statement of policy and priority-setting workshop notice. Federal Register 63:71542–71568.

Federal Register. 2000. Endangered and threatened wildlife and plants; Proposed designation of critical habitat for the California Red-legged Frog (*Rana aurora draytonii*). Federal Register 65:54892

Fellers GM, Drost CA. 1993. Disappearence of the Cascades frog *Rana cascadae* at the southern end of its range, California, USA. Biol Conserv 65:177–181.

Fellers GM, Green DE, Longcore JE. 2001. Oral chytridiomycosis in Mountain Yellow-Legged Frogs (*Rana muscosa*). Copeia 2001:945–953.

Gancedo B. Alonso-Gomez AL, de Pedro N, Corpas I, Delgado MJ, Alonso-Bedate M. 1995. Seasonal changes in thyroid activity in male and female frog, *Rana perezi*. Gen Comp Endocrinol 97:66–75.

Gancedo B, Corpas I, Alonso-Gomez AL, Delgado MJ, Morreale de Escobar G, Alonso-Bedate M. 1992. Corticotropin-releasing factor stimulates metamorphosis and increases thyroid hormone concentration in prometamorphic *Rana perezi* larvae. Gen Comp Endocrinol 87:6–13.

Goleman WL, Urquidi LJ, Anderson TA, Kendall RJ, Smith EE, Carr JA. 2002a. Environmentally relevant concentrations of ammonium perchlorate inhibit development and metamorphosis in *Xenopus laevis*. Environ Toxicol Chem 21:424–430.

Goleman WL, Carr JA, Anderson TA, 2002b. Environmentally relevant concentrations of ammonium perchlorate inhibit thyroid function and alter sex ratios in developing *Xenopus laevis*. Environ Toxicol Chem 21:590–597.

Goleman WL, Gentles BA, Hu F, Carr JA. 2004. Evidence that adult frogs are less responsive to the thyroid disrupting effects of ammonium perchlorate. Integ Comp Biol 43:1022.

Gupta BB, Chakrabarty P. 1990. Effects of thyroidal, gonadal and adrenal hormones on tissue respiration of streaked frog, *Rana limnocharis*, at low temperature. Indian J Exp Biol 28:23–26.

Gupta BB, Deka-Borah H. 1995. In vivo and in vitro effects of metabolic hormones on tissue respiration in toad, *Bufo melanostictus* during hibernation and active phase. Indian J Exp Biol 33:604–607.

Harrison KD, Zozzaro PE, Collie NL, Carr JA. 2002. Iodide transport in *Xenopus laevis* gut and skin. Amer Zool 41:1467.

Hauser KF, Gona AG. 1983. Effects of thyroidectomy and season on the external granular layer of the cerebellum in metamorphosing bullfrog tadpoles (*Rana catesbeiana*). Exp Neurol 79:265–277.

Hauser KF, Gona AG. 1984. Purkinje cell maturation in the frog cerebellum during thyroxine-induced metamorphosis. Neuroscience 11:139–155.

Hauser KF, Uray NJ, Gona AG. 1986. Granule cell development in the frog cerebellum during spontaneous and thyroxine-induced metamorphosis. J Comp Neurol 253:185–196.

Hayes TB. 1997. Hormonal mechanisms as developmental constraints on evolution: examples from the anura. Am Zool 37:482–490.

Hines H, Mahony M, McDonald K. 1999. An assessment of frog declines in wet subtropical Australia. In: Campbell A, editor. Declines and disappearances of Australian frogs. Canberra (AU): Environment Australia. p 44–63.

Hooth MJ, Deangelo AB, George MH, Gaillard ET, Travlos GS, Boorman GA, Wolf DC. 2001. Subchronic sodium chlorate exposure in drinking water results in a concentration-dependent increase in rat thyroid follicular cell hyperplasia. Toxicol Pathol 29:250–259.

Hoskins SG, Grobstein P. 1985. Development of the ipsilateral retinothalamic projection in the frog *Xenopus laevis*. III. The role of thyroxine. J Neurosci 5: 930–940.

Houlahan JE, Findlay CS, Schmidt BR, Meyer AH, Kuzmin SL. 2000. Quantitative evidence for global amphibian population declines. Nature 404:752–755.

Hu F, Gentles A, Goleman WL, Carr JA. 2003. Developmental stage and environmental iodide influence the antimetamorphic effects of perchlorate. Society of Environmental Toxicology and Chemistry 24th Annual Meeting; 2003 Nov 9–13; Austin, TX.

Ingram GJ, McDonald KR. 1993. An update on the decline of Queenslands frogs. In: Lunney D, Ayers D, editors. Herpetology in Australia: a diverse discipline. Mosman (AU): Royal Zoological Soc of NSW. p 297–303.

Kikuyama S, Miyakawa M, Arai Y. 1979. Influence of thyroid hormone on the development of preoptic-hypothalamic monoaminergic neurons in tadpoles of *Bufo bufo japonicus*. Cell Tissue Res 198:27–33.

Kinney KS, Felten SY, Cohen N. 1996. Sympathetic innervation of the amphibian spleen: developmental studies in *Xenopus laevis*. Dev Comp Immunol 20:l 51–59.

Krain LP, Denver RJ. 2004. Developmental expression and hormonal regulation of glucocorticoid and thyroid hormone receptors during metamorphosis in *Xenopus laevis*. J Endocrinol 181:91–104.

Lawler SP, Dritz D, Strange T, Holyoak M. 1999. Effects of introduced mosquitofish and bullfrogs on the threatened California red-legged frog. Cons Biol 13:613–622.

Manzon RG, Denver RJ. 2004. Regulation of pituitary thyrotropin gene expression during *Xenopus* metamorphosis: negative feedback is functional throughout metamorphosis. J Endocrinol 182:273–285.

May TW, Packer RK. 1976. Thyroid hormones stimulate in vivo oxygen consumption of adult *Rana pipiens berlandieri* at high environmental temperatures. Gen Comp Endocrinol 30: 525–527.

McDaniel LN. 2004. Ultraviolet radiation-induced genotoxicity in *Xenopus laevis* exposed to sodium perchlorate [MS thesis]. Lubbock (TX): Texas Tech Univ.

Miranda LA, Affanni JM, Paz DA. 2000. Corticotropin-releasing factor accelerates metamorphosis in *Bufo arenarum*: effect on pituitary ACTH and TSH cells. J Exp Zool 286:473–480.

Miranda LA, Dezi RE. 1997. Immunocytochemical distribution of corticotropin-releasing factor in the brain and hypophysis of larval *Bufo arenarum*; effect of KClO4 during early development. Tissue Cell 29:643–649.

Miranda LA, Pisano A, Casco V. 1996. Ultrastructural study on thyroid glands of *Bufo arenarum* larvae kept in potassium perchlorate solution. *Biocell* 20:147–153.

Morehouse EA, James TY, Ganley AR, Vilgalys R, Berger L, Murphy PJ, Longcore JE. 2003. Multilocus sequence typing suggests the chytrid pathogen of amphibians is a recently emerged clone. Mol Ecol 12:395–403.

Nieuwkoop PD, Faber J. 1994. Normal Table of *Xenopus laevis* (Daudin): A Systematical and Chronological Survey of the Development from the Fertilized Egg till the End of Metamorphosis, 3rd ed. Garland Publishing, New York, NY, USA.

Norris DO, Carr JA, Desan PH, Smock TK, Norman MF. 1992. Monoamines and their metabolites in the amphibian (*Ambystoma tigrinum*) brain: quantitative changes during metamorphosis and captivity. Comp Biochem Physiol 103A:279–283.

Okada R, Yamamoto K, Koda A, Ito Y, Hayashi H, Tanaka S, Hanaoka Y, Kikuyama S. 2004. Development of radioimmunoassay for bullfrog thyroid-stimulating hormone (TSH): effects of hypothalamic releasing hormones on the release of TSH from the pituitary in vitro. Gen Comp Endocrinol 135:42–50.

Prapunpoj P, Richardson SJ, Schreiber G. 2002. Crocodile transthyretin: structure, function, and evolution. Am J Physiol Regul Integr Comp Physiol 283:R885–896.

Prapunpoj P, Yamauchi K, Nishiyama N, Richardson SJ, Schreiber G. 2000. Evolution of structure, ontogeny of gene expression, and function of *Xenopus laevis* transthyretin. Am J Physiol Regul Integr Comp Physiol 279:R2026–2041.

Richter SC, Young JE, Johnson GN, Seigel RA. 2003. Stochastic variation in reproductive success of a rare frog, *Rana sevosa*: implications for conservation and for monitoring amphibian populations. Biol Conserv 111:171–177.

Robertson JC, Kelley DB. 1996. Thyroid hormone controls the onset of androgen sensitivity in the developing larynx of *Xenopus laevis*. Dev Biol 176:108–123.

Rollins-Smith LA, Blair P. 1990. Expression of class II major histocompatibility complex antigens on adult T cells in *Xenopus* is metamorphosis-dependent. Dev Immunol 1:97–104.

Rollins-Smith LA, Davis AT, Blair PJ. 1993. Effects of thyroid hormone deprivation on immunity in postmetamorphic frogs. Dev Comp Immunol 17:157–164.

Rollins-Smith LA, Flajnik MF, Blair PJ, Davis AT, Green WF. 1997. Involvement of thyroid hormones in the expression of MHC class I antigens during ontogeny in *Xenopus*. Dev Immunol 5:133–144.

Safi R, Bertrand S, Marchand O, Duffraisse M, de Luze A, Vanacker JM, Maraninchi M, Margotat A, Demeneix B, Laudet V. 2004. The axolotl (*Ambystoma mexicanum*), a neotenic amphibian, expresses functional thyroid hormone receptors. Endocrinology 145:760–772.

Schmidt CW. 1997. Amphibian deformities continue to puzzle researchers. Environ Sci Technol 31:324A–326A.

Shi YB. 2000. Amphibian metamorphosis: from morphology to molecular biology. New York (NY): Wiley-Liss.

Shi ZX, Levy A, Lightman SL. 1994. Thyroid hormone-mediated regulation of corticotropin-releasing hormone messenger ribonucleic acid in the rat. Endocrinology 134:1577–1580.

Smith PN, Theodorakis CW, Anderson TA, Kendall RJ. 2002. Preliminary assessment of perchlorate in ecological receptors at the Longhorn Army Ammunition Plant (LHAAP), Karnack, Texas. Ecotoxicology 10:305–313.

Sparling DW, Fellers GM, McConnell LL. 2001. Pesticides and amphibian population declines in California, USA. Environ Toxicol Chem 20:1591–1595.

Sparling DW, Harvey G, Nzengung V. 2003. Interaction between perchlorate and iodine in the metamorphosis of *Hyla versicolor*. In: Linder G, editor. Multiple stressor effects in relation to declining amphibian populations. West Conshohocken (PA): ASTM International. p 131–142. ASTM STP 1443.

Tasaki Y, Inoue M, Ishii S. 1986. Annual cycle of plasma thyroid hormone levels in the toad, *Bufo japonicus*. Gen Comp Endocrinol 62:404–410.

Theodorakis CW, Rinchard J, Park JW, McDaniel L, Liu F, Carr JA, Wages M. 2006. Thyroid endocrine disruption in stonerollers and cricket frogs from perchlorate-contaminated streams in east-central Texas. Ecotoxicology 15: 31-50.

Tietge JE, Holcombe GW, Flynn KM, Kosian PA, Korte JJ, Anderson LE, Wolf DC, Degitz SJ. 2005. Metamorphic inhibition of *Xenopus laevis* by sodium perchlorate: effects on development and thyroid histology. Environ Toxicol Chem 24:926–933.

[USACE] United States Army Corps of Engineers. 2004. Final Report: Bosque and Leon River Watersheds Study. Available from: http://www.swf.usace.army.mil/pubdata/ppmd/Perchlorate/deliverable final.html. Accessed 7/04.

[USFWS] US Fish and Wildlife Service. 1984. Houston toad recovery plan. Albuquerque (NM): USFWS. p 72.

[USFWS] US Fish and Wildlife Service. 2001. Wyoming toad (*Bufo baxteri*) population and habitat viability assessment (PHVA). Apple Valley (MN): p 109.

[USFWS] US Fish and Wildlife Service. 2003. Available from http://endangered.fws.gov/. Accessed May 2003.

[USGS] US Geological Survey. 2004. Available from http://www.usgs.gov/amphibian_faq.html. Accessed 5 Apr 2006.

United Nations Environment Programme World Conservation Monitoring Centre. 2006. Available from http://www.unep-wcmc.org. Accessed 4/2006.

Urbansky ET. 2002. Perchlorate as an environmental contaminant. Environ Sci Pollut Res Int 9:187–192.

Vandorpe G, Kuhn ER, Gevaerts H. 1990. Failure to relate thyroid hormones and in vitro 5'-monodeiodination activity to oocyte development and sex steroids in the giant swamp frog *Dicroglossus occipitalis* at the equator. Gen Comp Endocrinol 79:469–476.

Van Sande J, Massart C, Beauwens R, Schoutens A, Costagliola S, Dumont JE, Wolff J. 2003. Anion selectivity by the sodium iodide symporter. Endocrinology 144:247–252.

Vredenburg VT. 2004. Reversing introduced species effects: experimental removal of introduced fish leads to rapid recovery of a declining frog. Proc Natl Acad Sci 101:7646–7650.

West Plains IPM Update, 2002. Vol. 7, No. 16, Texas Cooperative Extension, Texas A&M System.

West Plains IPM Update. 2002. Vol. 7, No. 16. Available from: http://www.tpma.org/news_letters/_wplains_hockley/WPU09102002.htm. Accessed 5 Apr 2006.

World Resources Institute. 2006. Earthtrends: The Environmental Information Portal. Available from: http://earthtrends.wri.org/. Washington DC: World Resources Institute.

Yamauchi K, Eguchi R, Shimada N, Ishihara A. 2002. The effects of endocrine-disrupting chemicals on thyroid hormone binding to *Xenopus laevis* transthyretin and thyroid hormone receptor. Clin Chem Lab Med 40:1250–1256.

Yu L, Canas JE, Cobb GP, Jackson WA, Anderson TA. 2004. Uptake of perchlorate in terrestrial plants. Ecotoxicol Environ Saf 58:44–49.

seven

Perchlorate Effects in Fish

Christopher Theodorakis, Reynaldo Patiño, Erin M Snyder, Eric Albers

Introduction

Perchlorate anion is extremely water soluble and environmentally stable, resulting in rapid movement through ground and surface waters (Urbansky 2002). This fact, combined with the widespread use of perchlorate in industrial and military applications, indicates that the potential for perchlorate contamination in the aquatic environment, and thus the potential for exposure of fish to perchlorate, is significant. The number of known contaminated sites is probably limited by the small extent of monitoring for this contaminant. Furthermore, there is a paucity of knowledge concerning environmental fate, transport, toxicokinetics, bioavailability, and effects of this compound in aquatic systems.

Potential routes of entry of perchlorate into surface waters include

1) direct effluent discharge;
2) transfer from contaminated groundwater to the surface via seeps or springs, or use of contaminated groundwater for irrigation, with resultant runoff into surface waters;
3) percolation of water through contaminated soils; and
4) runoff from contaminated terrestrial ecosystems (Figure 7-1).

Furthermore, perchlorate may enter sediment pore water via surface water or ground water. Also, previous research has found that perchlorate accumulates in terrestrial plants (Smith et al. 2001) through uptake from soil, groundwater, or surface water, and most detritus in stream systems comes from fallen leaves of terrestrial plants in the riparian zone (i.e., streamside).

Perchlorate Ecotoxicology. Ronald J Kendall and Philip N Smith, editors.

ISBN 10: 1-880611-90-2   ISBN 13: 978-1-880611-90-6

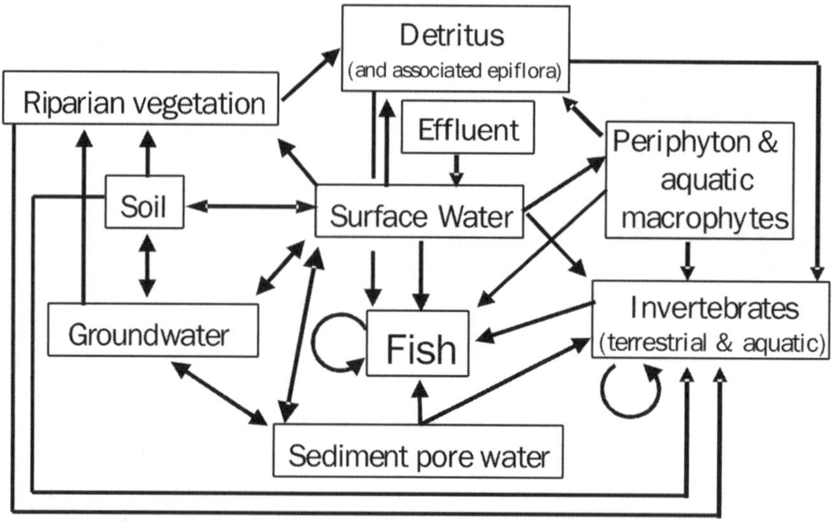

Figure 7-1 Possible routes of exposure of fish to perchlorate in aquatic ecosystems

(Such inputs of organic matter from terrestrial sources are known as "allochthonous inputs.") Thus, bioconcentration of perchlorate by terrestrial plants, followed by allochthonous inputs of detritus from such plants, is another possible pathway of entry of perchlorate into surface waters.

Fish may be exposed to perchlorate in contaminated surface waters via a number of different routes. A conceptual model of potential routes of uptake and transport through aquatic food webs is shown in Figure 7-1. First, because perchlorate is readily soluble, direct uptake from surface water through gills or permeable integument is probably a major route of entry. Another possible route of uptake in fish is through the food chain. Some fish feed on detritus, and so may be exposed to perchlorate in this fashion. The epiflora that lives upon detritus (saprophytic bacteria and fungi) could possibly take up perchlorate from the detritus or water. Perchlorate may also accumulate in periphyton, filamentous algae, and aquatic macrophytes (Anderson et al. 2004), representing a possible exposure pathway in herbivorous fish. Because many fish are invertebrivorous, any perchlorate taken up by aquatic invertebrates may be passed on to the fish. In addition, any fish feed upon terrestrial invertebrates that fall or are washed into the water, so this is an alternative pathway whereby terrestrial perchlorate in soil or plants may be transported to fish. Similar to fish, invertebrates may absorb perchlorate directly from contact with, or ingestion of, contaminated surface water, soil, sediment, or sediment pore water. Alternatively, invertebrates may be exposed to perchlorate from food items such as periphyton, detritus (or associated epiflora), or other invertebrates. Piscivorous fish may also be exposed to perchlorate in their prey. Although, for larger predacious fish (and for some large, aggressive invertebrates), there is the possibility of food chain transfer via consumption of amphibians (larvae or adults) and other small aquatic or semi-aquatic vertebrates; this is probably not a major exposure route. Whatever the route of exposure, the primary concern of perchlorate exposure is that it is known to affect thyroid function, causing subsequent hormone disruption and potential perturbations of metabolic physiological and developmental activities.

Although there is not as much information on thyroid biology in fish as there is in other vertebrates, it is known that thyroid hormones (THs) influence many aspects of fish physiology and biochemistry. In at least some species of teleosts, THs may have been reported to influence mitochondrial function, carbohydrate and lipid metabolism, growth, gonad and gamete development, embryo or larval development, metamorphosis, immune function, behavior, melanocyte development and skin pigmentation, larval metamorphosis, osmoregulation, neurogenesis, temperature acclimation, and steroidogenesis (Prosser et al. 1991; Soyano et al. 1992; Eales and Brown 1993; Leatherland 1994; Mylonas et al. 1994; Brown and Kim 1995; Cyr and Eales 1996; Tyler and Sumpter 1996; Gomez et al. 1997; Castonguay and Cyr 1998; Iwata et al. 1999; Schreiber and Specker 1999; Kumar et al. 2000; Subash Peter et al. 2000; Power et al. 2001; Varghese et al. 2001; Aas-Hansen et al. 2003; Brown et al. 2004). Disruption of TH synthesis and signaling during zebrafish (*Danio rerio*) larval development may also inhibit yolk sac resorption and development of the fins, gastrointestinal system, swim bladder, and lower jaw (Brown 1997; Liu and Chan 2002). In addition, teleost TH signaling pathways may regulate or interact with other endocrine or paracrine signaling pathways, including growth hormone, sex steroids, retinoids, corticosteroids, and prostaglandins (Cyr and Eales 1996; Gomez et al. 1997; Kim and Brown 1997; Kuhn et al. 1998; Sternberg and Moav 1999; van Anholt et al. 2003). However, broad generalities cannot be made because there is considerable interspecific variation. For example, in some species, THs may influence growth hormone and embryo or larval development, whereas in other species they may not (Nishioka et al. 1985; Luo and McKeown 1991; Tagawa and Hirano 1991; Melamed et al. 1995; Brown 1997; Liu and Chan 2002; Rousseau et al. 2002). Also, in some species gondadotropin-releasing hormone and gonadal or sexual maturation may be stimulated by THs, while in other species they may be inhibited by THs (Sullivan et. al 1989; Parhar et al. 2000). Besides these differences between species, interspecific variation, there may be intraspecific variation in TH levels and their effects on fish physiology. TH levels may vary within species or individuals as a function of season, photoperiod, diurnal periodicity, nutritional status, sex, and reproductive condition (MacKenzie et al. 1989; Pavlidis et al. 1991; Gomez et al. 1997; Leiner et al. 2000; Gaylord et al. 2001; Fontainhas-Fernandes et al. 2002; Aas-Hansen et al. 2003; Reddy and Leatherland 2003). Thus, thyroid-disrupting chemicals have the potential to perturb many aspects of fish physiology, but the effects of these chemicals on fitness and performance parameters remain uncertain (Brown et al. 2004).

While many aspects of thyroid anatomy and physiology are conserved between fish and tetrapods, there are some differences. First, follicles in most fish are not encapsulated within a discrete gland, with some notable exceptions such as tuna (*Thunnus* spp.) and parrotfish (Scaridae; Eales and Brown 1993, and references therein). In most teleosts, thyroid follicles are scattered throughout the basibranchial region (in or near the lower mandible, near the base of the gills) and around the bulbus arteriosus, but may also be present in the gills, liver, thymus, spleen, or kidney (Eales and Brown 1993; Leatherland 1994). Second, there may be differences between teleosts and tetrapods in terms of regulation of the hypothalamus–pituitary–thyroid axis. For example, in fish, thyrotropin-releasing hormone plays a lesser role (if any) in regulating thyroid-stimulating hormone (TSH) secretion. Other hormones may be involved in this regard, such as somatostatin (inhibitory) or corticotropic-releasing factor (stimulatory). These hormones are also transported from the hypothalamus to the pitutuary via direct innervation, rather than a blood portal system (Brown et al. 2004). Third, the primary thyroid-binding plasma proteins in fish may include albumin and lipoproteins, as well as transthyretin, and a percentage of TH may be transported in red blood cells in some species (Brown et al. 2004). Fourth, in some species such as trout, some T_3 may be sequestered in a slowly equilibrating pool such as skeletal muscle (Brown et al. 2004). Finally, fish obtain iodine for use in TH synthesis not only from the diet but also (and primarily) through absorption from the water via the gills (Brown et al. 2004). However, despite these differences between fish and tetrapods, there is a high degree of conservation among vertebrates in terms of physiological responses to thyroid-disrupting chemicals: Chemicals that disrupt TH, homeostasis in mammals do so in fish as well, via similar mechanisms (Leatherland 1994; Brown et al. 2001; Fox 2001).

Thus, when teleost fish are exposed to perchlorate, the effects on their thyroids are similar to those found in other vertebrates. The thyroidal effects of perchlorate are, in turn, affected by its uptake and toxicokinetics, and have the possibility to affect phyisiological and fitness parameters. The risk of perchlorate exposure to endemic fish populations may be assessing effects and toxicokinetics in the laboratory and the field, as well as by developing physiologically based models of the pharmacokinetics of perchlorate and TH kinetics in fish. This chapter summarizes current knowledge of perchlorate exposure and uptake in fish, lethal and sublethal effects, physiologically based pharmacokinetic (PBPK) models, and research needs.

Exposure

Uptake and toxicokinetics

There have been several studies on uptake and toxicokinetics of perchlorate in various fish species. These species include eastern mosquitofish (*Gambusia holbrooki*), zebrafish (*Danio rerio*), and bluegill sunfish (*Lepomis macrochirus*).

In one of these studies, female eastern mosquitofish were exposed to 0, 0.1, 1, 10, 100, and 1000 $mg \cdot L^{-1}$ sodium perchlorate for 2, 10, or 30 d, and then analyzed for perchlorate in whole-body homogenates (Bradford et al. 2006). No perchlorate was detected in mosquitofish exposed to the lowest concentrations of perchlorate (0, 0.1, and 1 $mg \cdot L^{-1}$ sodium perchlorate), for any exposure duration. Perchlorate was detected in mosquitofish exposed to 10, 100, and 1000 $mg \cdot L^{-1}$ sodium perchlorate, but the whole-body tissue concentrations were approximately 1/10 of the exposure concentration. Steady-state body burdens appeared to be reached relatively rapidly, within 2 d or less. The highest bioconcentration factors (BCFs) were in fish exposed to 10 and 100 $mg \cdot L^{-1}$ sodium perchlorate, with BCF, of 0.15 and 0.12, respectively. The BCF, for the 1000 $mg \cdot L^{-1}$ exposure was less than 0.01. An elimination experiment also was conducted in which fish were exposed to 100 $mg \cdot L^{-1}$ sodium perchlorate, and then allowed to depurate for 1, 2, 5, or 10 d. The uptake rate constant, elimination rate constant, and half-life of perchlorate in this species were 0.1 d^{-1}, 0.76 d^{-1}, and 0.91 d, respectively (Bradford et al. 2006).

Limited information on perchlorate uptake is also available for zebrafish. In a recent study, fish were exposed to ammonium perchlorate (AP)-derived perchlorate at measured concentrations of 677 $mg \cdot L^{-1}$ for 4 weeks or 18 $mg \cdot L^{-1}$ for 8 weeks (Patiño et al. 2003). The estimated BCFs in each case were 0.01 and 0.02, respectively.

In order to examine tissue distribution and tissue-specific uptake and elimination rates, Channel catfish (*Ictalurus punctatus*) were exposed to

100 mg·L^{-1} sodium perchlorate for 5 d, and allowed to depurate for 1, 2, 5, 10, and 20 d (Park 2003; Anderson et al. 2004). It was found that the BCFs were highest in the head and fillet. The fillet also had the lowest uptake rate constant and highest elimination rate constant, while the head and liver had the highest uptake rate constant and lowest elimination rate constant, respectively. These findings suggest that the head and fillet are the most important tissues to analyze when the toxicokinetics and uptake of perchlorate in fish are examined. Although perchlorate did accumulate in other tissues in the catfish (kidney, liver, gill, and gastrointestinal tract), these tissues had relatively low concentrations of perchlorate. Although perchlorate was rapidly eliminated from catfish tissues, it was detected in a few individuals even after 20 d of depuration. This finding suggests that depuration rates may vary greatly between individuals, and it has implications for interpretation of data from the field (see "Body burdens in field studies" below).

In another study, Dean et al. (2004) conducted bioconcentration tests on juvenile bluegill sunfish to 0, 1.5, and 15 mg·L^{-1} perchlorate anion (sodium salt) for 0, 1, 2, 3, 7, 14, 21, and 28 d. The BCF in this fish ranged from 0.73 L·kg^{-1} in the lowest treatment to 0.68 L·kg^{-1} in the highest treatment, with a geometric mean of 0.70 L·kg^{-1}.

In general, these results indicate that perchlorate is taken up by fish and reaches steady state relatively rapidly in some species, while it may take longer in others. Perchlorate was not detected in individuals exposed to environmentally relevant concentrations of perchlorate using ion chrmomatography (IC) with conductivity detection (mosquitofish, catfish, zebrafish), but for goldfish perchlorate detection was carried out using IC/MS/MS (ion chromatography/mass spectrometry/mass spectrometry). Differences between the catfish/mosquitofish vs. bluegill studies may represent species-specific differences or sensitivity of the techniques. Fish that were exposed to perchlorate in the laboratory did not bioaccumulate perchlorate because the tissue concentration of perchlorate was much lower than the exposure concentration of perchlorate. The uptake and elimination rate constants and BCFs may differ between species and between tissue types.

Body burdens in field samples

Environmental contamination of water sources occurs primarily near military and industrial installations where perchlorate is handled. To date, 14 of the United States, including Texas, have confirmed the presence of perchlorate in their ground and surface waters. In north-central and east Texas, perchlorate has been detected in water at 2 former military manufacturing sites, the Longhorn Army Ammunition Plant (LHAAP) (TNRCC 1999; Smith et al. 2001) and the Texas Naval Weapons Industrial Reserve Plant (TNWIRP) (Motzer 2001; Theodorakis et al. 2006a). At both sites, perchlorate was also measured in tissues of ecological receptors including algae, aquatic insects, and fish (Smith et al. 2001; Anderson et al. 2004).

Long Horn Army Ammunition Plant, Texas

In 1999, several areas within LHAAP, Karnack, Texas, were analyzed for perchlorate (Smith et al. 2001). Two perchlorate-contaminated sites were sampled for small fish species: Harrison Bayou and Goose Prairie Creek. Until the fall of 2002, Harrison Bayou received an effluent of treated groundwater that was contaminated with perchlorate. Goose Prairie Creek flows past an area where perchlorate was ground and processed (Building 25C), and where soil concentrations were as high as 322 $\mu g \cdot kg^{-1}$ (Smith et al. 2001). Perchlorate was detected in water samples from both sites at concentrations less than 85 $ug \cdot L^{-1}$. Perchlorate was detected only sporadically and in small, insectivorous or omnivorous fish species in the families Centrarchidae, Cyprinidae, Fundilidae, and Poeliciidae. When perchlorate was detected, concentrations in fish tissues exceeded those in the water.

Texas Naval Weapons Industrial Reserve Plant and Lakes Waco and Belton, Texas

In 2001 and 2002, potential perchlorate contamination was investigated in the Bosque River and Leon River watersheds, Texas, because AP was discovered outside the TNWIRP (former Hercules Plant), which lies within those 2 watersheds. Water samples were collected from 6 streams in and around the TNWIRP. Harris Creek, North Fork South Bosque River, South Fork South Bosque River, and Station Creek all originate on the TNWIRP. Wasp Creek does not originate on the Plant, but a portion of the creek runs through it. Coryell Creek neither originates on nor runs through TNWIRP. Perchlorate was detected in water samples from the 4 streams that originate on the TNWIRP. Two reservoirs formed by impoundment of the Bosque and Leon Rivers, Lake Waco and Lake Belton, respectively, were also sampled. Samples of fish fillet (to assess possible human exposure in edible portions) and head (because perchlorate concentrations in laboratory fish were greatest in the head) were collected for perchlorate analysis.

Results for the streams were similar to those for LHAAP. Perchlorate was detected sporadically, and when it was found, it usually occurred in cyprinids, centrarchids, and mosquitofish, although it also was detected in bullhead catfish (*Amerius spp.*). In particular, perchlorate was commonly found in green sunfish (*Lepomis cyanellus*). This finding is significant because the species is a popular sportfish and panfish in the area, colloquially known as "perch." Again, when perchlorate was detected, the concentration in biological tissues was much greater than in the water

A similar situation was observed for the fish sampled from Lakes Belton and Waco. Parts of the TNWIRP are located within the watersheds of both lakes, which receive inputs from streams originating on the TNWIRP. For these lakes, perchlorate was detected only sporadically in the fillets of several species of large fish, most commonly largemouth bass and channel catfish, although the concentration of perchlorate in the water was below the detection limit (1 $\mu g \cdot L^{-1}$) at the collection sites. These studies also found that the head had much higher concentrations of perchlorate than did the fillets, at least for black crappie.

Food chain transfer

In order to determine if perchlorate could be transferred through the food chain, Park et al. (2005) exposed bass to 500 $mg \cdot L^{-1}$ sodium perchlorate for 1, 10, or 30 d. There were 3 treatments for the bass: exposed to perchlorate via the food, via the water, or by a combination of food and water. The food consisted of fathead minnows exposed to 1000 $mg \cdot L^{-1}$ sodium perchlorate in the water for 2 d before they were fed to the bass. The body burden concentrations of perchlorate in largemouth bass indicated that food chain transfer of perchlorate occurred after 1 d of feeding. Moreover, fish exposed to perchlorate via food and water contained higher concentrations of perchlorate than did fish exposed to perchlorate in water only, although this was significant only at the 10-d exposure period. Two-way analysis of variance (ANOVA) indicated a significant interaction between the length of exposure and the mode of exposure ($P < 0.05$).

Effects

Acute toxicity

Several studies to date have found that perchlorate, either as sodium or ammonium salts, has low acute toxicity to fish. In one study, Patiño et al. (2003) found that the 5-d LC50 of AP to zebrafish embryos and larvae was 529 $mg \cdot L^{-1}$. Other studies using the sodium salt of perchlorate found 96-h LC50s of 404 and 1486 $mg \cdot L^{-1}$ for juvenile (<1 week old) eastern mosquitofish and 4-d old zebrafish, respectively (Liu et al. 2005; Park et al. 2006). Other species may be even less susceptible to perchlorate: the 96-h LC50s for rainbow trout (*Oncorhynchus mykiss*), fathead minnow (*Pimephales promelas*), and bluegill sunfish (*Lepomis macrochirus*) exposed to sodium perchlorate were 2100, 1655, and 1500 $mg \cdot L^{-1}$, respectively (Engineering, Science, and Technology 1998; Parsons et al. 2002). Dean et al. (2004) also calculated the 96-h LC50s of perchlorate for rainbow trout and bluegill as 2010 and 1470 $mg \cdot L^{-1}$, respectively, in close agreement with results from rainbow trout, fathead minnow, and bluegill, above. Chronic toxicity (survival endpoint) LC50, no-observed-efffects concentration (NOEC), and lowest-observed-effects concentration (LOEC), for fathead minnows were reported as 614, 155, and 280 $mg \cdot L^{-1}$ by Callahan et al. (1998). In general, it seems that the LC50s are very much higher than are the water concentrations seen in the field (Motzer 2001; Smith et al. 2001).

Sublethal Toxicity

Thyroid effects

Laboratory studies. Adult zebrafish exposed to AP at measured water perchlorate concentrations of 18 $mg \cdot L^{-1}$ for 8 weeks showed signs of thyroid follicle cell (nuclear) hypertrophy, hyperplasia, colloid depletion, and increased vascularization (angiogenesis) relative to control fish (Patiño et al. 2003). The degree of thyroid follicle vascularization in this study was particularly remarkable (Patiño et al. 2003), suggesting its possible utility as new biomarker of perchlorate exposure. In a follow-up study (Mukhi et al. 2005), subadult (3-month-old) zebrafish were exposed to AP-derived perchlorate at measured concentrations of 0, 11, 90, 1131, and 11 480 $\mu g \cdot L^{-1}$ for up to 12 weeks and allowed to recover in clean water for another 12 weeks. The LOEC of perchlorate that induced angiogenesis after 2 weeks of exposure was 90 $\mu g \cdot L^{-1}$, whereas hypertrophy was not detected at this time. By 12 weeks of exposure, the LOEC for angiogenesis remained at 90 $\mu g \cdot L^{-1}$, whereas the LOEC for hypertrophy was 1131 $\mu g \cdot L^{-1}$ (Mukhi et al. 2005). The content and distribution of thyroxine (T_4)-like immunoreactivity of the colloid in histological sections of thyroid follicles was also examined by immunocytochemistry. The thyroid follicles displayed a "colloidal T_4 ring" previously observed at the interphase between the follicular epithelium and the lumen, as reported in larval zebrafish (Wendl et al. 2002). The intensity of this colloidal T_4 ring was reduced by exposure to perchlorate, with a 2-week LOEC, of 1131 $\mu g \cdot L^{-1}$ and a 12-week LOEC of 11 $\mu g \cdot L^{-1}$ (Mukhi et al. 2005). Whole-body T_4 concentrations (measured by radioimmunoassay) were not affected by perchlorate in this study. Thus, a change in the intensity of the colloidal T_4 ring following relatively extended periods of exposure (12 weeks) was the most sensitive of the parameters of thyroid toxicity condition examined. All perchlorate-induced changes in thyroid follicles were reversible, but residual effects on colloidal T_4 ring intensity as well as angiogenesis were still present after 12 weeks of recovery in the fish previously exposed to the highest concentration of perchlorate (11480 $\mu g \cdot L^{-1}$). Thus, the sensitivity and longevity of changes in colloidal T_4 ring intensity and angiogenesis suggest their usefulness as novel markers of perchlorate exposure and toxicity thyroid effects in fish.

Effects of perchlorate on mosquitofish thyroid disruption also have been examined by Bradford et al. (2005). In this particular study, adult female eastern mosquitofish were exposed to 0, 0.1, 1, 10, 100, and 1000 $mg \cdot L^{-1}$ (nominal concentrations) sodium perchlorate for 2, 10, and 30 d in the laboratory. Thyroid follicle histopathology was examined in stained sections of the head, and whole-body T_4 concentrations were determined by radioimmunoassay. An increase in follicular epithelial height was noted, particularly in fish exposed to perchlorate for 30 d. In addition, an increase in the occurrence of hyperplasia, hypertrophy, and colloid depletion also was observed in a dose- and time-dependent fashion. There was a small decrease in T_4 concentrations in fish exposed to perchlorate for 30 d, but this effect was less pronounced than the effects on thyroid histopathology.

Additionally, adult male and female common goldfish (*Carassius auratus*) were exposed separately to environmentally relevant concentrations of sodium perchlorate (14 to 31 000 $\mu g \cdot L^{-1}$) in a laboratory flow-through exposure system (Crouch 2004). Treatments consisted of combinations of perchlorate concentration (measured anion concentration) and exposure duration as follows:

- control (<4 $\mu g \cdot L^{-1}$), 30 d;
- 14 $\mu g \cdot L^{-1}$, 30 d;
- 130 $\mu g \cdot L^{-1}$, 30 d;
- 31 000 $\mu g \cdot L^{-1}$-30 d;
- control (<4 $\mu g \cdot L^{-1}$), 60 d;
- and 31 000 $\mu g \cdot L^{-1}$, 60 d.

Both thyroid and reproductive endpoints were examined. A new, subjective, nonparametric rank-order assessment method was developed to assign activity scores to thyroid tissue in the subpharyngeal region (pharyngeal thyroid, lower jaw) and head kidney based on epithelial cell height, colloid depletion, number of follicles affected, and severity of effects (Crouch 2004). A digital imaging system was used to measure thyroid follicle epithelial cell heights in the same tissues to allow a comparison between the new subjective assessment method and the more traditional and objective measurement method. Thyroid activity scores resulting from the assessment method were more sensitive to the effects of perchlorate than were epithelial cell height measurements. A significant increase in pharyngeal thyroid activity occurred in males exposed to 31 000 $\mu g \cdot L^{-1}$ for 60 d and in females exposed to 31 000 $\mu g \cdot L^{-1}$ for 30 d or 60 d. A, significant increase in head kidney thyroid activity occurred in males exposed to 31 000 $\mu g \cdot L^{-1}$ for 60 d and in females exposed to 1200 $\mu g \cdot L^{-1}$ for 30 d or to 31 000 $\mu g \cdot L^{-1}$ for 30 d or 60 d (Crouch 2004).

Field studies. Field studies on the effects of perchlorate are generally lacking. In studies by Theodorakis et al. (2006a) and Anderson et al. (2004), central stonerollers (*Campostoma anomalum*) and green sunfish (*Lepomis cyanellus*) were collected from the streams in and around TNWIRP. A reference site (Coryell Creek) was also chosen. Streams were sampled in October 2001 and March 2002. In October, Coryell, Wasp, Station, and Harris Creeks were sampled, while in March, the South Bosque River replaced Harris Creek as a sampling site. Histopathological indices, including the percentage of thyroid follicles that were hyperplastic or hypertrophic or that exhibited colloid depletion, were determined for stonerollers. The percentage of follicles that exhibited angiogenesis (extent of vascular tissue within the follicles), which has also been identified as an indicator of thyrotoxicity (Patiño et al. 2003), was measured (Anderson et al. 2004). The epithelial cell heights were determined for green sunfish collected in March (Anderson et al. 2004). In addition, the percent of individuals that exhibited evidence of angiogenesis in thyroidal tissue was also evaluated for fish collected in October.

Overall, the level of effect was associated with proximity to the TNWIRP. Fish collected from streams that originated on the TNWIRP grounds (Harris Creek and Station Creek) were impacted more than were fish collected from other streams. The Wasp Creek and South Bosque stonerollers seemed to be affected to a level intermediate between the level of effect observed in fish from the other streams that originate on TNWIRP and the level of effect in fish from Coryell Creek. This finding is consistent with the fact that Wasp Creek and the South Basque River are in closer proximity to TNWIRP than is Coryell Creek, and that Wasp Creek runs through TNWIRP while Coryell Creek does not. Furthermore, the trend in detectable perchlorate concentrations in fish among streams parallels the evidence of thyroid disruption. Wasp Creek was originally thought to be a reference site, but perchlorate was found in some of the fish from Wasp Creek; however, the incidence of detectable perchlorate in fish was less in Wasp Creek than in the streams that originate on TNWIRP. Overall, researchers concluded that the data collected from this study suggested that perchlorate is negatively impacting thyroid homeostasis of the fish in streams that originate in or run through TNWIRP.

The data from the October 2001 sampling indicate the following:

- For thyroid follicle epithelial heights, the level of effect is in the order of Harris Creek > Wasp Creek > Station Creek > Coryell Creek.
- For percentage occurrence of hyperplasic follicles, the relative levels of effect were Station Creek > Wasp Creek = Harrison Creek > Coryell Creek.
- For percent occurrence of follicles with depleted colloid, the levels of effect were Harrison Creek = Station Creek > Wasp Creek = Coryell Creek.
- For angiogenesis, the levels of efffect were Station Creek > Wasp Creek > Harris Creek > Coryell Creek.

For the stonerollers collected in March, no statistically significant differences in the occurrence of hyperplasia were found, while the trends for occurrence of depleted colloid and epithelial cell height were Station = South Bosque = Wasp > Coryell and Coryell = South Bosque > Wasp = Station Creek. For occurrence of angiogenic responses, the level of effect was Wasp=Station > Coryell = Harris. Green sunfish from Coryell Creek had the smallest thyroid follicles, fish from the South Bosque River had the largest, and the other 2 sites had intermediate levels (Anderson et al. 2004; Theodorakis et al. 2006b). A comparison of thyroid epithelial cell heights between green sunfish and stonerollers suggests that the green sunfish in the South Bosque River are impacted more than stonerollers from the same river (Anderson et al. 2004).

Macrophage aggregates

Macrophage aggregates (MAs) or centers are generally considered to serve in the recycling and removal of cellular debris; the sequestering, destruction, and removal of cellular toxicants; antigen trapping and presentation to lymphocytes; and other functions (Tsujii and Seno 1990; Wolke 1992; Agius and Roberts 2003). In fish, exposure to environmental contaminants is known to increase the incidence of MAs in some tissues, and this response may indicate the occurrence of tissue injury or immunotoxic response (Weeks et al. 1992; Wolke 1992; Fournie et al. 2001; Agius and Roberts 2003). A recent study with zebrafish and eastern mosquitofish examined the histological effects of perchlorate on the trunk kidney, which in teleosts serves excretory and hemopoietic functions (Capps et al. 2005). Adult zebrafish of both sexes were exposed in the laboratory to waterborne, AP-derived perchlorate at measured concentrations of 18 $mg \cdot L^{-1}$ for 8 weeks. Adult male mosquitofish were exposed to waterborne sodium perchlorate at measured perchlorate concentrations of 1 to 92 $mg \cdot L^{-1}$ for 8 weeks. Control fish were kept in untreated water. Histological analyses showed that MAs were present in the hemopoietic region of the trunk kidney in both species (the head kidney was not examined). The estimated percent area of trunk kidney sections occupied by MAs was higher in zebrafish exposed to perchlorate at 18 $mg \cdot L^{-1}$ when compared to controls.

In male mosquitofish, the incidence of renal MAs increased proportionally with sodium perchlorate concentration and was significantly different from controls at 92 $mg \cdot L^{-1}$. These observations confirm that the kidney is affected by exposure to perchlorate in fish. The concentrations at which perchlorate had clear effects on the incidence of renal MAs (18 $mg \cdot L^{-1}$ in zebrafish and 92 $mg \cdot L^{-1}$ in mosquitofish) are at the high end of perchlorate concentrations previously reported in the environment; perchlorate concentrations of up to 33 $mg \cdot L^{-1}$ have been reported in some contaminated aquatic habitats (Smith et al. 2001). Thus, the environmental relevance of these findings is uncertain at the present time. However, it has been reported that the magnitude of the thyroidal effects of perchlorate is a function not only of exposure concentration but also of exposure duration (Fernandez Rodriguez et al. 1991; Mukhi et al. 2005). Thus, if the thyroid system were involved in the mechanism of perchlorate-induced kidney MAs, exposures to perchlorate at lower concentrations but for longer periods of time than those used in the present study may also affect the structure and function of the fish kidney.

Fitness parameters: growth, reproduction, development

A recent review of the effects of contaminants on the thyroid system concluded that xenobiotic-induced changes in thyroid function have yet to be clearly linked to impairments in the fitness or survival of fish (Brown et al. 2004). In agreement with this conclusion, the data presently available for fish have failed to demonstrate clear effects of perchlorate on parameters such as growth and reproduction at environmentally relevant concentrations.

Zebrafish. An 8-week exposure to AP-derived perchlorate at high environmentally relevant levels (18 $mg \cdot L^{-1}$) did not affect the reproductive performance (spawn volume and fertilization rate) of adult zebrafish, although it greatly disrupted their thyroid histological condition (Patiño et al. 2003). Also, the growth and pubertal development of subadult zebrafish were not affected by exposure to AP-derived perchlorate (0 to 11 480 μg·L^{-1}) during or after an experimental exposure period of 12 weeks (Mukhi et al. submitted). Results similar to those obtained with zebra-fish (Patiño et al. 2003; Mukhi et al. submitted) were reported in a study with rats, where a 2-generation exposure to AP did not affect reproductive performance or embryonic development (York et al. 2001).

Mosquitofish. Park et al. (2006) examined the effects of perchlorate on reproduction and growth of eastern mosquitofish exposed to sodium perchlorate. This species is commonly found at perchlorate-contaminated sites such as TNWIRP and LHAAP (Smith et al. 2001; Theodorakis et. al. 2006a). Adult and 7-d-old fry were exposed to sodium perchlorate at 0, 1, 10, and 100 mg·L^{-1} (nominal concentrations). Effects on growth and reproduction were determined after 14 and 56 d of exposure for fry and adults, respectively. Their results indicated that perchlorate exposure had a stimulatory effect on fecundity, gonadal somatic index (GSI = [gonad weight / body weight] *100), and average egg or embryo mass, at least for some treatments. Growth was also stimulated at 1 mg·L^{-1} but was inhibited at 10 mg·L^{-1}. These results fail to provide evidence that perchlorate inhibits growth and reproduction in eastern mosquitofish at environmentally relevant concentrations.

Goldfish. Adult male and female common goldfish were exposed separately to environmentally relevant concentrations (14 to 31 000 μg·L^{-1}) of sodium perchlorate for 30 to 60 d in a laboratory flow-through exposure system under conditions designed to stimulate spring gonadal recrudescence (Crouch 2004). Perchlorate exposure did not affect GSI in either sex, and there were no differences in gonad development, as assessed histologically, among treatment groups. However, the fish were not exposed during the period of gonadal recrudescence most sensitive to impairment of gonadal growth. Evaluations of plasma sex steroid concentrations 17ß-estradiol, testosterone, and 11-ketotestosterone) and TH concentrations (T_4 and T_3) are underway.

Fathead minnows. In chronic, 35-d early life-stage toxicity studies, no effects of perchlorate were found in fathead minnows at perchlorate ion concentrations up to 490 mg·L^{-1} perchlorate anion (Engineering, Science, and Technology 2000).

Lamprey. The effects of potassium perchlorate at very high concentrations (100 mg·L^{-1}) have also been investigated in sea lamprey (*Petromyzon marinus*) from a landlocked population. Contrary to most teleosts, THs inhibit larval sea lamprey metamorphosis (Holmes and Youson 1993).

It was found that perchlorate treatment enhanced larval metamorphosis (Kao et al. 1999). Potassium perchlorate was also found to reduce serum TH concentrations and to affect lipid metabolism in this species (Youson et al. 1995; Kao et al. 1999). Similar effects were seen for American brook lamprey (*Lampetra appendix*; Holmes et al. 1999). Because of the high concentrations of perchlorate used in these studies, the environmental relevance of their findings remains uncertain.

Toxicity of perchlorate mixtures

Perchlorate and arsenate

Thyroid hormones have been found to influence levels of antioxidant enzymes and oxidative stress in fish (Chugh and Singh 1991; Varghese et al. 2001). Thus, thyroidal status may affect an organism's ability to respond to oxidative stress, or may determine the relative toxicity of pro-oxidant contaminants. In addition, perchlorate is a powerful oxidant, and so may itself induce oxidative stress in biological systems. Thus, there may be interaction between perchlorate and pro-oxidant compounds in inducing toxic responses. This hypothesis was tested by Liu et al. (2005), who exposed early-stage zebrafish larvae to sodium arsenate and sodium perchlorate. Arsenate was chosen because it is a model pro-oxidant that may co-occur with perchlorate in contaminated sites, and because several investigators (Lopez-Torrez et al. 2000, Das and Chainy 2001, Rahaman et al. 2001) have found that thyroid status modulates oxidative stress by arsenate in laboratory animals. Liu et al. (2005) found evidence that arsenate and perchlorate exhibited strictly additive toxicity, possibly because of similar, overlapping, or compensatory modes of action (e.g., they both may be pro-oxidants and thyroid-disrupting compounds).

Perchlorate and octylphenol

Recently, Cruz-Li (2004) completed a study examining the effects of perchlorate, octylphenol (a xenoestrogen), and their mixture on mortality and reproduction in zebrafish. For the mortality test, fish were exposed to 20 $\mu g \cdot L^{-1}$ AP and 3 $\mu g \cdot L^{-1}$ octylphenol (OP, nominal concentration), and a mixture of both during the embryo stage (0 to 3 d post fertilization [PF]), the larval stage (3 to 6 d PF), or the embryo–larval period (0 to 6 d PF). Mortality was noted during the exposure period and up to 30 d PF, (i.e., 24 to 27 d after exposure was ended). For all treatments (AP, OP, AP+OP), mortality was low (<20%) during the exposure period, but increased to approximately 60% to 80% after 20 d PF. Control mortality was approximately 50% to 60% at this time. For the fish that were exposed during the embryo and larval stages, no difference was found among the treatments, but mortality was greater in treated fish than in control fish, especially at 20 d PF. This finding indicated that all treatments induced delayed mortality long after exposure ceased. For fish exposed during the embryo-larval period, the relative mortality increased in the order of control = AP < OP < AP+OP. Among the exposure regimes, the fish exposed during the entire embryo–larval period were least sensitive to AP, while fish exposed during the larval period were least sensitive to the mixture. For the reproduction tests, fish were exposed as above and allowed to reach sexual maturity in clean water (to test for latent effects). There were no differences among treatments for egg production, egg size, and fertilization success.

Predictive Physiologically Based Pharmacokinetic Models

Toxicokinetic models

A physiologically based toxicokinetic (PBTK) model was developed to simulate the movement of perchlorate within channel catfish (*Ictalurus punctatus*) organs (Albers 2004). Contaminant movement was governed by a series of mass-balanced differential equations programmed in Matlab. Model compartments and blood flow can be seen in Figure 7-2. General equations used in the model were taken from PBTKs developed by Nichols et al. for rainbow trout (1990, 1991) and channel catfish (1993). The rate of change for each of the 7 compartments (skin, kidney, muscle, GI tract, liver, gill, and thyroid) at individual time steps was obtained by solving the differential equations simultaneously using numerical integration. Fish gill physiology was kept biologically accurate by accounting for countercurrent chemical flux, including both flow and diffusion limitations (Erickson and McKim 1990b). Flow-limited distribution was assumed, that is, chemical equilibrium existed between the tissues and blood leaving the compartment. Additionally, portal blood flow was incorporated into the kidney and liver from poorly perfused tissue and richly perfused tissue, respectively. This represents the first and only PBTK model of perchlorate in fish.

Physiological parameters for blood flow and cardiac output in the channel catfish were obtained from existing literature (Hughes 1984; Erickson and McKim 1990a, 1990b; Nichols et al. 1990; McKim et al. 1994). Tissue blood flow was scaled by body weight to account for differences in literature and laboratory fish sizes. Tissue masses were measured in undosed juvenile channel catfish (Bradford 2002; Park 2003), with their

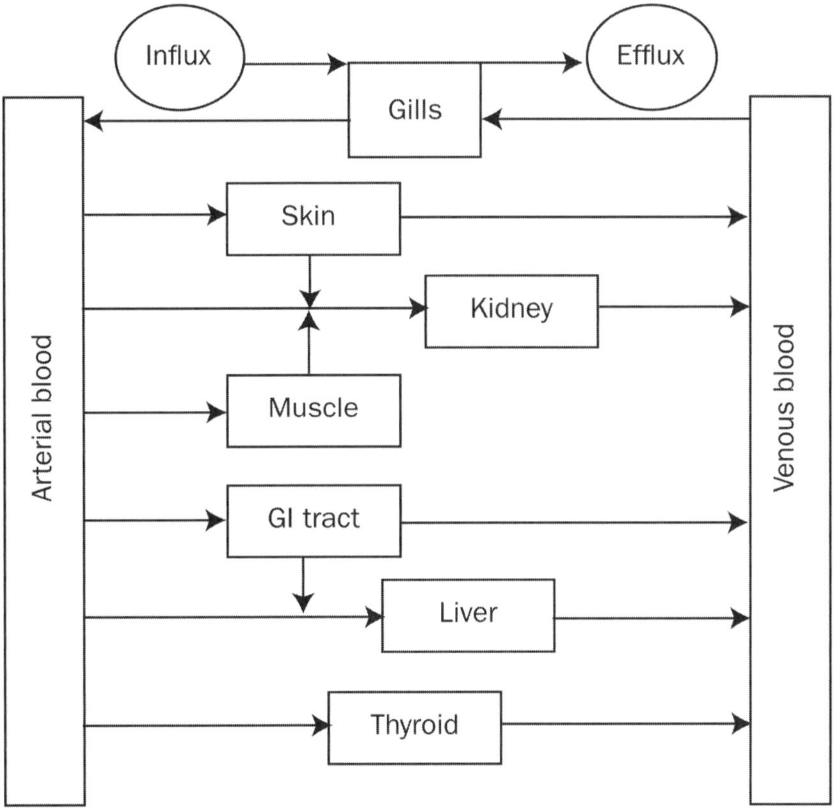

Figure 7-2 Flow diagram of the PBTK model for perchlorate inhalation in fish

averages used to determine their fractional mass. An equation for chemical flux at the gills (Erickson and McKim 1990b) was used to simulate uptake through inspired water. This method accounts for counter-current exchange with flows separated by a diffusion barrier made up of the gill epithelium and stagnant boundary layers in adjacent blood and water channels (Nichols et al. 1993). The PBTK model was calibrated using data from a 10-d elimination study in channel catfish (Park 2003) discussed previously (see "Uptake and toxicokinetics" above). By fitting the model to the elimination data, as well as to the uptake data, we were able to account for the lack of a urinary excretion term (Figure 7-3).

Thyroid hormone models

A 6-compartment model of combined T_3 and T_4 kinetics (Figure 7-4) originally developed for mammals (DiStefano 1986; Hershman et al. 1986; Bianchi et al. 1987; Pilo et al. 1990) and subsequently applied to rainbow trout (Sefkow et al. 1996) was used to simulate the secretion of

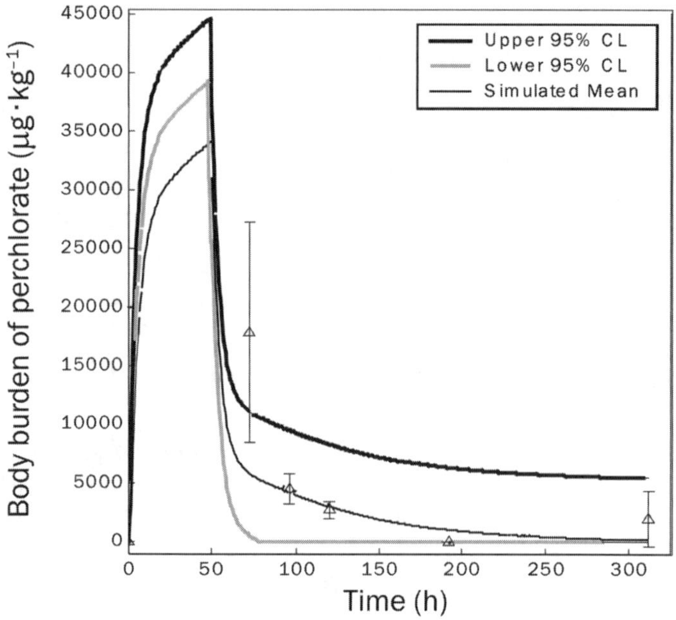

Figure 7-3 Calibration curve for thyroid compartment using elimination data

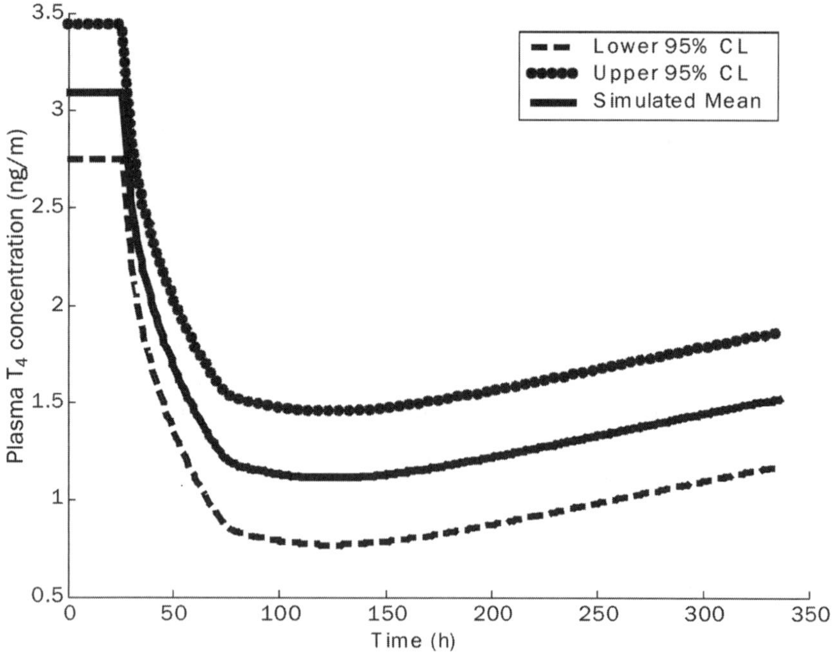

Figure 7-4 Simulated perchlorate-induced thyroid hormone inhibition in the T_4 plasma compartment based on thyroid tissue concentrations from Park (2003) 10-d elimination study

T_3 and T_4 in channel catfish, as well as the impact of perchlorate on the T_4 secretion rate (Albers 2004). The 6 compartments included

1) T_4 slow-exchange tissue pool,
2) T_4 rapid-exchange tissue pool,
3) T4 plasma pool,
4) T_3 slow-exchange tissue pool,
5) T_3 rapid-exchange tissue pool, and
6) T_3 plasma pool.

Skeletal muscle makes up the majority of the slow-exchange tissue pool. The rapid-exchange tissue pool is made up primarily of the liver but also of the kidney and gill, all of which possess deiodinating capacity (MacLatchy and Eales 1992; Eales and Brown 1993). Model parameters were reported by Sefkow et al. (1996) as either uniquely identifiable or determined within a range using the method of interval identifiability analysis (DiStefano 1983). Steady-state equations were developed by Sefkow et al. (1996) based on data from fasted rainbow trout injected with radiolabeled T_3 and T_4. Mass-balance ratios were used to calibrate the model, with plasma T_3 and T_4 compartments adjusted to meet levels reported by Gaylord et al. (2001) in channel catfish.

This model is a fairly simplistic representation of TH regulation in fish. It does not simulate the feedback loop and the related levels of TSH or corticotropin-releasing factor (CRF). The most notable drawback to this model is the lack of hormone pools. Fish, like mammals, possess stored TH pools that can be released even if iodide uptake is inhibited by perchlorate. As a result of these pools, there may be no observable effect of perchlorate exposure for some time, until the pool is depleted (Patiño et al. 2003). Until the nature of this pool is quantified, however, we are not able to incorporate that feature into the model.

T_4 hormone inhibition by perchlorate was calculated from data on mosquitofish dosed with sodium perchlorate for 2, 10, or 30 d at doses of 0, 0.1, 1, 10, 100, and 1000 $mg \cdot L^{-1}$ (Bradford et al. 2005). Whole-body T_4 concentrations were determined by radioimmunoassay for pooled groups of fish. The inhibition term was applied to the three T_4 compartments in the model (plasma, rapid exchange, and slow exchange) through adjustment of the T_4 secretion rate term.

The model was applied to the 10-d elimination study scenario to test its ability to rebound once perchlorate is removed from the system (Figure 7-4). The three T_4 compartments exhibited the greatest decrease in hormone levels (~51%) with a concomitant decrease in the T_3 compartments

(~16%) as expected. We saw a rapid decrease in hormone levels during exposure, followed by a gradual increase. Because the thyroid model is initiated with the simulated thyroid tissue concentrations from the PBTK model, hormone levels fluctuate with thyroid perchlorate concentrations. Evidence of this relationship can be seen in the plots of hormone concentrations that are the inverted shape of the plots of tissue data. While hormone recovery is evident once perchlorate exposure ceases, we do not know if the recovery rate is accurate.

In general, however, the modeled results do not coincide with actual results in other species to date. In the model, T_4 concentrations decrease by more than 3-fold within 3 d of exposure to perchlorate (Figure 7-4). In other fish species tested to date, much longer chronic exposures are needed to effect such reductions, and the effects are not this dramatic (Bradford et al. 2005). This result may be due to the lack of a model to accommodate TSH, feedback loops, or colloid storage of thyroglobulin. Until these issues are resolved, however, the utility of this model remains uncertain.

Data Gaps and Future Research Needs

Although there has been considerable research on the biological effects of perchlorate, there are still many gaps in the knowledge of its ecological risk. For example, more information is needed regarding the effects of perchlorate on fitness parameters. Although no adverse effects of environmentally relevant concentrations of perchlorate on growth or reproduction have been demonstrated in any fish species, the available data are insufficient to make generalizations, especially for the early life stages that are typically more sensitive to disruption by environmental contaminants. Accordingly, early life-stage and full life-cycle or partial life-cycle tests are necessary for adequate analyses of risk assessment. In mammals, TH disruption also has been implicated in male reproductive effects, including effects on sperm production, testicular function, and steroid synthesis (Buzzard et al. 2003; Krassas and Perros 2003; Maran 2003), but such effects have not been examined in fish.

Although it is known that perchlorate may affect the amount of THs produced by adult fish, the potential for perchlorate to cause trans-generational effects in fish has yet to be investigated. Embryonic fish require TH for development but generally do not synthesize the hormone themselves until after hatching. THs in embryos are supplied via maternal transfer to the eggs. If thyroid synthesis is disrupted in the mother, it may affect her ability to transfer THs to her eggs and thus affect embryonic development. Although perchlorate may affect the amount of THs produced by adult fish, especially after prolonged exposures, the effects of perchlorate-induced thyroid endocrine disruption on maternal transfer of THs to eggs, and potential consequent effects on fish embryo development, are matters that have yet to be resolved.

Another potential area for future research is examination of alternative physiological fitness endpoints of perchlorate exposure. For example, embryonic neuronal development is controlled by THs (Denver et al. 1997); thus, it is possible that perchlorate-induced thyroid disruption could result in neurobehavioral effects, but this has not been investigated. In addition, perchlorate exposure and thyroid suppression have been found to suppress the immune system in amphibians (Rollins-Smith 1993; Weetman 1994), and THs have been found to affect immune function in fish as well (Harris and Bird 2000), but the potential effect of perchlorate on fish immune function has not been explored. Finally, possible effects of environmentally relevant concentrations of perchlorate on other thyroid-responsive physiological endpoints (e.g., lipid metabolism, temperature acclimation, osmoregulation, pigmentation, mitochondrial function, embryo or larval development) and expression of thyroid-responsive and fitness-related genes merit further attention. In particular, studies on thyroid-responsive gene expression could be used to provide

1) a method for distinguishing between pathological effects and compensatory responses and
2) tools for studying thyroid disruption in embryos, larvae, and early juveniles, which are too small for conventional TH assays.

In a related light, studies that examine interactions between THs and other endocrine-, exocrine-, and paracrine-signaling pathways are an important consideration. For example, because growth hormone gene expression may be controlled or influenced by TH status, does perchlorate affect growth hormone levels? Other potential effects could be less direct via cross talk between TH and sex hormone, corticosteroid-, and retinoid-signaling pathways.

An additional data gap that needs to be addressed is the interaction between perchlorate and naturally occurring and anthropogenic chemicals. The most relevant naturally occurring chemical is iodide or other oxidative states of iodine (e.g., iodate) because availability of iodine in the water could be critical for determining the magnitude of perchlorate-induced effects. In terms of anthropogenic chemicals, it is not known if perchlorate affects the toxicity, environmental chemistry and fate, toxicokinetics, or detoxification of other toxicants, and vice versa. Potential mechanisms for such interactive effects could include redox reactions between chemicals, thyroid-mediated changes in metabolism of chemicals, or nonthyroidal effects on fish physiology. Finally, the toxicity and behavior of perchlorate in complex mixtures is unknown.

In terms of PBPK and TH models, more work needs to be done on the testing, validation, refinement, and sensitivity analysis of the models. In particular, inclusion of components for tissue- or compartment-specific peripheral deiodination, metabolism, excretion, and storage (especially storage in follicle colloids) would greatly enhance utility, realism, and accuracy of thyroid models. The models could also be expanded to include molecular, receptor- or ligand-based, toxicodynamic, and computational approaches.

Finally, more field studies need to be conducted to properly assess the ecotoxicological risk of perchlorate. These may include microcosm or mesocosm studies or field surveys to further characterize environmental fate of perchlorate and pathways of exposure for organisms in aquatic systems. Such studies would provide answers to questions that are critical for assessing ecological risk of perchlorate. For example, to what extent do each of the potential pathways and routes of exposure play in the exposure and effects of perchlorate in fish? Also, what are the possible indirect effects of perchlorate on fish populations? Examples of indirect effects include

1) perturbations of aquatic food webs resulting from toxic effects on plankton or macroinvertebrates,
2) differential sensitivities to perchlorate among species, and
3) interactions between perchlorate and natural variables or stressors (e.g., pathogens, salinity, pH, temperature, dissolved oxygen, suspended sediment, ultraviolet radiation, food availability).

Other manipulative or survey-type studies are needed to determine the extent of perchlorate exposure on reproduction and development in field situations, and the effects of perchlorate contamination on populations or communities of fish. Until these issues are resolved, it will be difficult to perform a complete ecological risk assessment of the effects of perchlorate in natural environments.

Conclusions

Research to date indicates that there is little risk to native fish populations due to acute toxic effects of perchlorate, but that chronic exposure to perchlorate in the laboratory and the field can result in structural and physiological perturbations to the fish thyroid gland tissue and its hormones. Perchlorate does not bioconcentrate in fish tissues in laboratory settings, but measurement of water concentrations alone is insufficient to predict body burdens or toxicological effects in field situations. This discrepancy between field and laboratory studies may be due to complex multiple exposure pathways and/or highly variable perchlorate water concentrations in time and space. At the present time, there is no evidence of perchlorate-induced effects such as growth, survival, and reproduction at environmentally relevant concentrations. However, further studies are needed before sound risk assessments of perchlorate are possible. Full life-cycle and multigenerational studies will be necessary to fully understand the ecological effects of perchlorate on individuals and populations of fish. The models developed and under development have potential applications for calculating probability exposure or effects distributions for use in probabilistic risk assessments (see Chapter 10). To date, evidence for detrimental effects of perchlorate on ecologically relevant fitness parameters at environmentally relevant concentrations remains equivocal. However, because of the limited number of species tested, limited number and types of chronic tests, paucity of information on behavior and toxicity of perchlorate in complex mixtures, and the fact that THs play a part in so many critical aspects of fish physiology, further studies are needed before comprehensive risk assessments of perchlorate can be carried out.

References

Aas-Hansen O, Johnsen HK, Vijayan MM, Jorgensen EH. 2003. Development of seawater tolerance and concurrent hormonal changes in fed and fasted Arctic charr at two temperature regimes. Aquaculture 222:135–148.

Agius C, Roberts RJ. 2003. Melano-macrophage centres and their role in fish pathology. J Fish Disease 26:499–509.

Albers EP. 2004. Using High Performance Computing and Visualization to Enhance Risk Assessment Methodology: A Case Study With Perchlorate. Ph.D. Dissertation, Texas Tech University, Lubbock, TX.

Anderson TA, Smith PN, McMurry ST, Carr JA, Theodorakis CW, Jackson WA, Dixon KR. 2004. Ecological risk assessment of ammonium perchlorate on fish, amphibians, and mammals in the Lake Belton and Lake Waco watersheds. An integrated laboratory and field investigation. In: US Army Corps of Engineers, Bosque and Leon River Watershed Study, Final Report. US Army Corps of Engineers Fort Worth District, Fort Worth, TX.

Bianchi T, Iervasi G, Pilo A, Vitek F, Ferdeghini M, Cazzuola F, Giraudi G. 1987. Role of serum carrier proteins in the peripheral metabolism and tissue distribution of thyroid hormones is familial dysalbuminemic hyperthyroxinemia and congenital elevation of thyroxine-binding globulin. J Clin Invest 80:522–534.

Bradford CM, Carr JA, Rinchard J, Theodorakis C. 2005. Perchlorate affects thyroid function in eastern mosquitofish (*Gambusia holbrooki*) at environmentally relevant concentrations. Eviron Sci Technol 39:5190–5195.

Bradford CM, Park J-W, Rinchard J, Anderson TA, Liu F, Theodorakis CW. 2006. Uptake and elimination of perchlorate in eastern mosquitofish. Chemosphere 63:1591–1597.

Brown CL, Kim BG. 1995. Combined application of cortisol and triiodothyronine in the culture of larval marine finfish. Aquaculture 35:79–86.

Brown DD. 1997. The role of thyroid hormone in zebrafish and axolotl development. Proc Natl Acad Sci 94:13011–13016.

Brown RP, Greer RD, Mihaich EM, Guiney PD. 2001. A critical review of the scientific literature on potential endocrine-mediated effects in fish and wildlife. Ecotox Environ Saf 49:17–25.

Brown SB, Adams BA, Cyr DG, Eales JG. 2004. Contaminant effects on the teleost fish thyroid. Environ Toxicol Chem 23:1680–1701.

Buzzard JJ, Wreford NG, Morrison JR. 2003. Thyroid hormone, retinoic acid, and testosterone suppress proliferation and induce markers of differentiation in cultured rat Sertoli cells. Endocrinology 144: 3722–3731

Callahan C, Sprenger M, Long GC, Porter RC. 1998. Perchlorate ecological risk studies - a report on literature reviews and studies conducted by the ecological impact/transport and transformation subcommittee of the Interagency Perchlorate Steering Committee. Institute for Environment Safety and Occupational Health Risk Analysis, Brooks, AFB TX.

Capps T, Mukhi S, Rinchard J, Theodorakis CW, Blazer VS, Patiño R. 2005. Exposure to perchlorate induces the formation of macrophage aggregates in the trunk kidney of zebrafish and mosquitofish. J Aquat Anim Health 6:145–151.

Castonguay M, Cyr DG. 1998. Effects on temperature on spontaneous and thyroxine-stimulated locomotor activity of Atlantic cod. J Fish Biol 53:303–313.

Chugh SP, Singh S. 1991. Role of thyroid-hormones in rotenone treated teleost - *Anabas-testudineus bloch* - effect on oxidative enzyme-activities. Am Zool 31:A71.

Crouch NT. 2004. Investigation of the effects of perchlorate on thyroid and reproductive system function in goldfish [MS thesis]. State College (PA): Pennsylvania State Univ.

Cruz-Li EI. 2004. Effects of ammonium perchlorate, 4(tert-octyl) phenol and their mixture on zebrafish (*Danio rerio*) [PhD dissertation]. Lubbock (TX): Texas Tech Univ.

Cyr DG, Eales JG. 1996. Interrelationships between thyroidal and reproductive endocrine systems in fish. Rev Fish Biol Fish 6:165–200.

Das K, Chainy GBN. 2001. Modulation of rat liver mitochondrial antioxidant defence system by thyroid hormone. Biochim Biophys Acta-Mol Basis Dis 1537:1–13.

Dean KE, Palacheck RM, Noel JM, Warbritton R, Aufderheide J, Wireman J. 2004. Development of freshwater water-quality criteria for perchlorate. Environ Toxicol Chem 23:1441–1451.

Denver RJ, Pavgi S, ShiY-B. 1997. Thyroid hormone-dependent gene expression program for Xenopus neural development. J Biol Chem 272: 8177–8188.

DiStefano III JJ. 1983. Complete parameter bounds and quasi-identifiability conditions for a class of unidentifiable linear systems. Math Biosci 65:51–68.

DiStefano III JJ. 1986. Modeling approaches and models of the distribution and disposal of thyroid hormones. In: Hennemann G, editor. Thyroid hormone metabolism. New York (NY): Dekker.

Eales JG, Brown SB. 1993. Measurement and regulation of thyroidal status in teleost fish. Rev Fish Biol 3:299–347.

Engineering, Science, and Technology. 1998. Results of acute and chronic toxicity testing with sodium perchlorate. Report 2900. Brooks Air Force Base, TX, USA.

Engineering, Science, and Technology. 2000. Results of chronic toxicity testing with sodium perchlorate using *Hyalella azteca* and *Pimephales promelas*. Report 3505. Brooks Air Force Base, TX, USA.

Erickson RJ, McKim JM. 1990a. A simple flow-limited model for exchange of organic chemicals at fish gills. Environ Toxicol Chem 9:159–165.

Erickson RJ, McKim JM. 1990b. A model for exchange of organic chemicals at fish gills:Flow and diffusion limitations. Aquatic Toxicol 18:175–198.

Fernandez Rodriguez A, Galera Davidson H, Salguero Villadiego M, Moreno Fernandez A, Martin Lacave I, Fernandez Sanz J. 1991. Induction of thyroid proliferative changes in rats treated with antithyroid compound. Anat Histol Embryol 20:289–298.

Fontainhas-Fernandes A, Gomes E, Reis-Henriques MA, Coimbra J. 2002. Plasma thyroid hormones and hepatic nucleic acids in relation to sex of tilapia Oreochromis niloticus. J Appl Ichthyol 18:185 191.

Fournie JW, Summers JK, Courtney LA, Engle VD, Blazer VS. 2001. Utility of macrophage aggregates as an indicator of fish exposure to degraded environments. J Aquat Animal Health 13:105-116.

Fox GA. 2001. Effects of endocrine disrupting chemicals on wildlife in Canada: past, present and future. Water Qual Res J Can 36:233–251.

Gaylord TG, MacKenzie DS, Gatlin III DM. 2001. Growth performance, body composition and plasma thyroid hormone status of channel catfish (*Ictalurus punctatus*) in response to short-term feed deprivation and refeeding. Fish Physiol Biochem 24:73–79.

Gomez JM, Boujard T, Boeuf G, Solari A, LeBail PY. 1997. Individual diurnal plasma profiles of thyroid hormones in rainbow trout (*Oncorhynchus mykiss*) in relation to cortisol, growth hormone, and growth rate. Gen Comp Endocrinol 107:74–83.

Harris J, Bird DJ. 2000. Modulation of the fish immune system by hormones. Vet Immunol Immunopathol 77: 163–176.

Hershman J, Nadamanee K, Sugawara M, Pekary AE, Ross R, Singh B, DiStefano III JJ. 1986. Thyroxine and triiodothyronine kinetics in cardiac patterns taking amiodarone. Acta Endocrinol 111:193–199.

Holmes J, Youson JH. 1993. Induction of metamorphosis in landlocked sea lampreys, Petromyzon marinus. J Exp Zool 267:598–604

Holmes JA, Chu H, Khanam SA, Manzon RG, Youson JH. 1999. Spontaneous and induced metamorphosis in the American brook lamprey *Lambpetra appendix*. Can J Zool 77:959–971.

Hughes GH, 1984. General anatomy of the gills. In: Hoar WS, Randall DJ, editors. Fish physiology, Vol. X; Gills, Part A, Anatomy, gas exchange and acid-base regulation. New York (NY): Academic Pr. p 1–71.

Iwata M, Hutchison MJ, Watanabe T, Yamashita T, Watanabe Y, Kasai N, Tsuboi H, Satoh R, Yamada H, Chiba H. 1999. Regulation mechanisms of the downstream migratory behaviours in salmonids fishes. Bull Tohoku Natl Fish Res Inst, Spec. Issue. no. 62, Part 3. Anadromous Fish Control. p 117–131.

Kao YH, Manzon RG, Sheridan MA, Youson JH. 1999. Study of the relationship between thyroid hormones and lipid metabolism during KClO4- induced metamorphosis of landlocked lamprey, *Petromyzon marinus*. Comp Biochem Physiol 122C:363–373.

Kim BG, Brown CL. 1997. Interaction of cortisol and thyroid hormone in the larval developmental of Pacific threadfin. Am Zool 37:470–481.

Krassas GE, Perros P. 2003. Thyroid disease and male reproductive function. J Endocrinol Invest 26: 372–380.

Kuhn ER, Geris KL, van der Geyten S, Mol KA, Darras VM. 1998. Inhibition and activation of the thyroidal axis by the adrenal axis in vertebrates. Comp Biochem Physiol A. 120:169–174.

Kumar RS, Ijiri J, Shigeho K, Trant JM. 2000. Changes in the expression of genes encoding steroidogenic enzymes in the channel catfish (*Ictalurus punctatus*) ovary throughout a reproductive cycle. Biol Reprod 63:1676–1682.

Leatherland JF. 1994. Reflections on the thyroidology of fishes:from molecules to humankind. Guelph Icthyol Rev 2:3–67.

Leiner KA, Han GS, Mackenzie DS. 2000. The effects of photoperiod and feeding on the diurnal rhythm of circulating thyroid hormones in the red drum, *Sciaenops ocellatus*. Gen Comp Endocrinol 120:88–98.

Liu F, Kendall RJ, Theodorakis CW. 2005. Joint toxicity of sodium arsenate and sodium perchlorate to zebrafish *Danio rerio* larvae. Environ Toxciol Chem 24:1505–1507.

Liu YW, Chan WK. 2002. Thyroid hormones are important for embryonic to larval transitory phase in zebrafish. Differentiation 70:36–45.

Lopez-Torres M, Romero M, Barja G. 2000. Effect of thyroid hormones on mitochondrial oxygen free radical production and DNA, oxidative damage in the rat heart. Mol Cell Endocrinol 168:127–134.

Luo D, McKeown BA. 1991. The effect of thyroid hormone and glucocordicoids on carp growth hormone-releasing factor (GRF)-induced growth hormone (GH) release in rainbow trout (*Oncorhynchus mykiss*). Comp Biochem Physiol A. 99:621–626.

MacKenzie DS, Thomas P, Farrar SM. 1989. Seasonal changes in thyroid and reproductive steroid hormones in female channel catfish (*Ictalurus punctatus*) in pond culture. Aquaculture 78:63–80.

MacLatchy DL, Eales JG. 1992. Properties of T4 5'-deiodinating systems in various tissues of the rainbow trout, *Oncorhynchus mykiss*. Gen Comp Endocrinol 86:313–322.

Maran, RRM. 2003. Thyroid hormones: their role in testicular steroidogenesis. Arch Androl 49: 375–388

McKim JM, Nichols JW, Lien GJ, Bertelsen SL. 1994. Respiratory-cardiovascular physiology and chloroethane gill flux in the channel catfish (*Ictalurus punctatus*). J Fish Biol 44:527–547.

Melamed P, Eliahu N, Levavi-Sivan B, Ofir M, Farchi-Pisanty O, Rentier-Delrue F, Smal J, Yaron Z, Naor Z. 1995. Hypothalamic and thyroid regulation of growth hormone in tilapia. Gen Comp Endocrinol 97:13–30.

Motzer WE. 2001. Perchlorate: problems, detection, and solutions. Environ Forensic 2:301–311.

Mukhi S, Carr JA, Anderson TA, Patino R. 2005. Development and validation of new biomarkers of perchlorate exposure in fishes. Environ Toxicol Chem 24:1107–1115.

Mylonas CC, Sullivan CV, Hinshaw JM. 1994. Thyroid-hormones in brown trout (*Salmo trutta*) reproduction and early development. Fish Physiol Biochem 13:485–493.

Nichols JW, McKim JM, Anderson ME, Gargas ML, Clewell HJ III, Erickson RJ. 1990. A physiologically based toxicokinetics model for the uptake and disposition of waterborne organic chemicals in fish. Toxicol Appl Pharmacol 106:433–447.

Nichols JW, McKim JM, Anderson ME, Lien GJ, Hoffman AD, Bertelson SL. 1991. Physiologically based toxicokinetics modeling of three waterborne chloroethanes in rainbow trout (*Oncorhyncus mykiss*). Toxicol Appl Pharmacol 110:374–389.

Nichols JW, McKim JM, Lien GJ, Hoffman AD, Bertelsen SL, Gallinat CA. 1993. Physiologically-based toxicokinetic modeling of three waterborne chloroethanes in channel catfish, *Ictalurus punctatus*. Aquat Toxicol 27:83–112.

Nishioka RS, Grau EG, Bern HA. 1985. In vitro release of growth hormone from the pituitary gland of tilapia, *Oreochromis mossambicus*. Gen Comp Endocrinol 60:90–94.

Parhar IS, Soga T, Sakuma Y. 2000. Thyroid hormone and estrogen regulate brain region-specific messenger ribonucleic acids encoding three gonadotropin-releasing hormone genes in sexually immature male fish, *Oreochromis niloticus*. Endocrinology 141:1618–1626.

Park J-W. 2003. Toxicological effects and kinetics of perchlorate on fishes [MS thesis. Lubbock (TX): Texas Tech Univ.

Park J-W, Rinchard J, Anderson TA, Liu F, Theodorakis CW. 2005. Food chain transfer of perchlorate in largemouth bass, *Micropterus salmoides*. Bull Environ Toxicol Chem 74:56–63.

Park J-W, Rinchard J, Liu F, Anderson TA, Kendall RJ, Theodorakis CW. 2006. The thyroid endocrine disruptor perchlorate affects reproduction, growth, and survival of mosquitofish. Ecotox Environ Saf 63:343–352

Patiño R, Wainscott MR, Cruz-Li EI, Balakrishnan S, McMurry C, Blazer VS, Anderson TA. 2003. Effects of ammonium perchlorate on the reproductive performance and thyroid follicle histology of zebrafish. Environ Toxicol Chem 22:1115–1121.

Pavlidis M, Dessypris A, Christofidis I. 1991. Seasonal fuctuations in plasma thyroid hormones, in two strains of rainbow trout (*Oncorhynchus mykiss*), during the first reproductive and second reproductive cycle: relation with their photoperiodically altered spawning time. Aquaculture 99:365–385.

Pilo A, Iervasi G, Vitek F, Fereghini M, Cazzuola F, Bianchi R. 1990. Thyroidal and peripheral production of 3,5,3'-triiodothyronine in humans by multicompartmental analysis. Am J Physiol 258:E715–E726.

Power DM, Llewellyn L, Faustino M, Nowell MA, Bjornsson BT, Einarsdottir IE, Canario AVM, Sweeney GE. 2001. Thyroid hormones in growth and development of fish. Comp Biochem Physiol C 130:447–459.

Prosser CL, Graham G, Galton V. 1991. Hormonal-regulation of temperature-acclimation in catfish hepatocytes. J Comp Physiol B 161:117–124.

Rahaman SO, Ghosh S, Mohanakumar KP, Das S, Sarkar PK. 2001. Hypothyroidism in the developing rat brain is associated with marked oxidative stress and aberrant intraneuronal accumulation of neurofilaments. Neurosci Res 40:273–279.

Reddy PK, Leatherland JF. 2003. Influences of photoperiod and alternate days of feeding on plasma growth hormone and thyroid hormone levels in juvenile rainbow trout. J Fish Biol 63:197–212.

Rollins-Smith LA, Davis AT, Blair PJ. 1993. Effects of thyroid hormone deprivation on immunity in postmetamorphic frogs. Dev Comp Immunol 17:157–164.

Rousseau K, Le Belle N, Sbaihi M, Marchelidon J, Schmitz M, 2002. Evidence for a negative feedback in the control of eel growth hormone by thyroid hormone. J Endocrinol 175:605–613.

Schreiber A, Specker J. 1999. Early larval development and metamorphosis in the summer flounder:changes in per cent whole-body water content and effects of altered thyroid status. J Fish Biol 55:148–157.

Sefkow AJ, Distefano JJ, III, Himick BA, Brown SB, Eales JG. 1996. Kinetic analysis of thyroid hormone secretion and interconversion in the 3-day fasted rainbow trout, *Oncorhynchus mykiss*. Gen Comp Endocrinol 101:123–138.

Smith PN, Theodorakis CW, Anderson TA, Kendall RJ. 2001. Preliminary assessment of perchlorate in ecological receptors at the Longhorn Army Ammunition Plant (LHAAP), Karnack, Texas. Ecotoxicology 10:305–313.

Soyano K, Saito T, Nagae M, Yamauchi K. 1992. Effects of thyroid hormone on gonadotropin-induced steroid production in medaka, *Oryzias latipes*, ovarian follicles. Fish Physiol Biochem 11:265–272.

Sternberg H, Moav B. 1999. Regulation of the growth hormone gene by fish thyroid/retinoid receptors. Fish Physiol Biochem 20:331–339.

Subash Peter MC, Lock RA, Wendelaar Bonga SE. 2000. Evidence for an osmoregulatory role of thyroid hormones in the freshwater Mozambique tilapia *Oreochromis mossambicus*. Gen Comp Endocrinol 120:157–167.

Sullivan CV, Bernard MG, Hara A, Dickhoff WW. 1989. Thyroid hormones in trout reproduction: enhancement of gonadotropin-releasing hormone analogue and partially purified salmon gonadotropin-induced ovarian maturation in vivo and in vitro. J Exp Zool 20:188–195.

Tagawa M, Hirano T. 1991. Effects of thyroid-hormone deficiency in eggs on early development of the medaka, *Oryzias latipes*. J Exp Zool 257:360–366.

Theodorakis CW, Rinchard J, Anderson T, Liu F, Park J-W, Costa F, McDaniel L, Kendall R, Waters A. 2006a. Perchlorate in fish from a contaminated site in east-central Texas. Environ Pollut 139:59–69.

Theodorakis CW, Rinchard J, Carr JA, Park J-W, McDaniel L, Liu F, Wages M. 2006b. Thyroid endocrine disruption in stonerollers and cricket frogs from perchlorate-contaminated streams in east-central Texas. Ecotoxicology 15:31–50.

Tsujii T, Seno S. 1990. Melano-macrophage centers in the aglomerular kidney of the sea horse (teleosts): Morphologic studies on its formation and possible function. Anat Rec 226:460–470.

Tyler CR, Sumpter JP. 1996. Oocyte growth and development in teleosts. Rev Fish Biol Fish 6:287–318.

Urbansky, ET. 2002. Perchlorate as an environnemental contaminant. Environ Sci Pollut Res Int 9: 187–192.

van Anholt RD, Spanings T, Koven W, Bonga SEW. 2003. Effects of acetylsalicylic acid treatment on thyroid hormones, prolactins, and the stress response of tilapia (*Oreochromis mossambicus*). Am J Physiol-Reg I 285: R1098–R1106.

Varghese S, Shameena B, Oommen OV. 2001. Thyroid hormones regulate lipid peroxidation and antioxidant enzyme activities in *Anabas testudineus* (Bloch). Comp Biochem Physiol B 128:165–171.

Weeks BA, Anderson DP, DeFour AP, Fairbrother A, Govern AJ, Lahvis GP, Peters G. 1992. Immunological biomarkers to assess environmental stress. In: Huggett RJ, Kimerle RA, Mehrle Jr PM, Bergman HL, editors. Biomarkers, physiological and histological markers of anthropogenic stress. Boca Raton (FL): Lewis. p 211–234.

Weetman, AP. 1994. The immunomodulatory effects of antithyroid drugs. Thyroid 4:145–146.

Wendl T, Lun K, Mione M, Favor J, Brand M, Wilson SW, Rohr KB. 2002. Pax2.1 is required for the development of thyroid follicles in zebrafish. Development 129:3751–3760.

Wolke RE. 1992. Piscine macrophage aggregates:a review. Ann Rev Fish Dis 2:91–108.

York RG, Brown WR, Girard MF, Dollarhide JS. 2001. Two-generation reproduction study of ammonium perchlorate in drinking water in rats evaluates thyroid toxicity. Int J Toxicol 20:183–197.

Youson JH, Holmes JA, Leatherland JF. 1995. Serum concentrations of thyroid hormones in $KClO_4$-treated larval sea Lampreys (*Petromyzon marinus* L.). Comp Biochem Physiol C 111:265–270.

eight

Perchlorate Effects on Invertebrates

Todd A Anderson, Mindy Landrum, Erin M Snyder

Introduction

Concerns have been raised that perchlorate released into the environment could disrupt the normal production of thyroid hormones needed to regulate metabolism, growth, and development in humans. Despite the human health assessments, relatively little is understood about potential ecological exposure and effects from environmental perchlorate contamination. Because thyroid hormones have analogous functions in other species, fish and wildlife exposed to environmental perchlorate could potentially suffer disruptions in endocrine homeostasis (Kendall et al. 1998). Most forms of perchlorate in the environment exist as the free anion in solution. Monitoring data indicate that perchlorate contamination is not confined to surface water and/or groundwater; perchlorate does occur as a contaminant in soils and sediments (Smith et al. 2001, 2004; Tan et al. 2004). Soils and sediments have very little anion exchange capacity; therefore, perchlorate absorbs weakly to most minerals. Consequently, perchlorate is highly mobile in soil and sediment and biologically available for uptake into soil and sediment-dwelling organisms.

Because of their limited mobility, earthworms and other invertebrates are appropriate species for monitoring the impact of contaminants and changes in soil and sediment structure (Paoletti 1999). Aquatic and terrestrial invertebrates make a major contribution to animal biomass and play a significant role in trophic transfer of nutrients (and contaminants). Several vertebrate and invertebrate species feed upon these lower-trophic-level organisms (Abdul Rida and Bouche 1995). Thus, they become an additional exposure pathway to the numerous predators that consume them if contaminants accumulate in invertebrate tissues.

Perchlorate Ecotoxicology. Ronald J Kendall and Philip N Smith, editors.

ISBN-10: 1-880611-90-2 ISBN-13: 978-1-880611-90-6

Earthworms

Earthworms are important in the later stages of soil formation and in maintaining soil structure and fertility (Edwards and Lofty 1977). These organisms contribute to soil productivity by incorporating decaying organic matter into soil and turning over this organic matter and mixing it with other soil fractions (Edwards 1983). Earthworms also aid in improving soil aeration, drainage, and moisture-holding capacity. As worms burrow, they increase soil macroporosity and aeration. These pores and the burrows themselves improve drainage while simultaneously improving soil moisture capacity as capillary water around the finer soil particles is retained more efficiently. Because of their importance in soil conservation, earthworms can be utilized as a tool to evaluate different transformations and impacts. Agricultural activities such as plowing, various tillage operations, fertilizing, and application of chemical pesticides have dramatic influence on these animals. As such, earthworms are an appropriate species for monitoring the potential impact of contaminants and changes in soil structure and agricultural practices (Paoletti 1999). Earthworms are capable of absorbing chemicals through respiratory exchange by diffusion through the skin, which is closely underlain by capillary networks. The capillaries allow the circulating blood to attain oxygen and eliminate carbon dioxide through the moist body surface (Gaddie and Douglas 1975).

Ecologically, earthworms are near the bottom of the terrestrial trophic food chain and have many predators, such as birds, moles, and even some large mammals. Earthworms not only are sensitive to chemicals (see following sections) but have a tendency to concentrate compounds such as organochlorine insecticides and heavy metals into their tissues (Edwards 1983). These chemicals may have little effect on the earthworms directly but may be taken up into predator tissues and affect other animals higher in terrestrial food chains.

In addition to their key roles in soil fertility and trophic transfer, earthworms are common in a wide variety of soils, in both temperate areas and the tropics. They are relatively large in size and easy to handle. Earthworms can be readily collected and identified. From data collected in laboratory and field studies, earthworms are known to be affected by a number of organic and inorganic chemicals, which may also be taken up into their tissues and concentrated. These organisms are easily bred in the laboratory for toxicity testing, and their longevity makes it unlikely that numerous worms would die during the period of a toxicity test in untreated media (OECD 1984).

Earthworms are hermaphroditic, each individual having complete male and female sexual organs. This trait is thought to be an adaptation to sedentary life in which there are relatively few contacts between individuals (Gaddie and Douglas 1975). During copulation, a mutual exchange of sperm occurs. Thus, from one mating it is possible for each earthworm to produce offspring.

When two individuals mate, they come into juxtaposition and overlap each other with anterior ends facing in opposite directions. Each earthworm then secretes mucus forming a thick band about the clitellar regions of their bodies. The clitellum is a swollen ring formed by the thickening of epithelium, which contains many gland cells. The mucous bands surround both bodies, binding the copulating individuals tightly together. Each earthworm then dispenses seminal fluid that is directed along the grooves to the seminal receptacles of the other earthworm where it is then stored. Once sperm has been deposited in each earthworm's seminal receptacles, the pair separates and departs (Gaddie and Douglas 1975).

Seminal receptacles function as sperm storage. Thus, copulation in earthworms does not directly lead to fertilization. The sperm and egg cells are released and mature in the coelom in special pouches. The sperm mature in the seminal vesicles, and the egg cells mature in ovisacs. The cocoon-laying process begins when the gland cells of the clitellum secrete a mucous ring that glides forward over the body of the earthworm. As the ring passes the oviduct opening, it receives several mature eggs. Similarly, as the ring passes over the more anterior sperm receptacles, it receives sperm. Fertilization of the eggs occurs in the mucous ring. The mucous ring gradually slips off over the earthworm's anterior end and closes up, forming a cocoon that contains a fluid albuminous nutriment and several zygotes, which will be deposited in the soil until the small earthworms hatch (Gaddie and Douglas 1975).

Earthworm cocoons are greenish-yellow when first laid but turn darker as they age. When they are ready to hatch, the cocoons turn a dark brown or purple. The formation of young earthworms in the cocoon and the number produced by each cocoon depend on the number of eggs and spermatozoa deposited into the cocoon as it passed over the body of the earthworm. A well-fed earthworm in favorable surroundings (loose bedding) is likely to produce a cocoon containing more eggs and spermatozoa than an earthworm that is actively searching for food or moving to a more comfortable environment (Gaddie and Douglas 1975). The embryos develop directly into young earthworms within the cocoon. As previously stated, the cocoon contains an albuminous nutriment upon which the embryos feed. Under normal conditions, the newly developed embryos deplete the nutriment within 14 to 20 days and then must emerge from the cocoon to find another source of nourishment. Emergent earthworms, except for size, are fully developed and self-sufficient (Gaddie and Douglas 1975).

Because reproduction may be inhibited or halted at contaminant concentrations far below a given LC50, toxicity is often not a useful indicator of the potential ecological consequence of a contaminant (Neuhauser et al. 1985). However, reproduction is a sensitive parameter and is important for the preservation of animal populations. Moreover, evaluations of earthworm growth and reproduction following exposure to chemicals provide insight into the potential sublethal effects of compounds. Studies of this kind involve exposing the test organism to a chemical that has been incorporated into some sort of artificial soil substrate.

Aquatic invertebrates

Aquatic invertebrates are ecologically important animals due to their trophic interactions and effects on aquatic ecosystems. Macroinvertebrates are major primary and secondary consumers, feeding on coarse and fine particulate matter, periphyton, and animals (Benke 1996). By decomposing allochthonous material, they serve as the fundamental link between organic matter resources they consume and the fish that prey upon them. They are the primary food resources for fish and other vertebrates, including recreationally and commercially important species, and are crucial first foods for many fish species. Freshwater macroinvertebrates are ubiquitous, diverse, and abundant. For example, in many lotic systems, the macroinvertebrate community (excluding the meiofauna) consists of several hundred species representing many phyla (Hauer and Resh 1996). Meiofauna, which are smaller than macroinvertebrates and include the young of many macroinvertebrate species, constitute an even larger proportion of the benthic (or bottom-dwelling) invertebrate community in terms of numbers and species richness and represent more than 95% of the benthic community in most streams.

Aquatic invertebrates are known to concentrate certain contaminants from the environment. They may accumulate contaminants from the water column by direct uptake of water-soluble contaminants from the water or by filter-feeding or otherwise ingesting contaminated items (Barron 2003). Benthic invertebrates also might be exposed by ingesting contaminated prey items or sediment particles to which contaminants are bound, by adsorption to or direct uptake across body surfaces in contact with sediment, or by direct uptake from sediment pore water or from the boundary layer of water overlying sediment.

Invertebrates can influence the distribution of contaminants in aquatic ecosystems, thus affecting their own exposure and the exposure of other organisms. Vertical mixing of sediments by invertebrates can increase or decrease removal, transformation, and bioavailability of sediment-associated contaminants. Aquatic invertebrates can concentrate bioaccumulative contaminants from environmental media, and accumulation of persistent, hydrophobic contaminants by benthic organisms and subsequent consumption by predators is an important pathway for such contaminants to enter aquatic food webs.

Pelagic and sediment-dwelling invertebrates are also important in the development of water quality criteria for environmental contaminants. Lists and descriptions of freshwater toxicity tests developed for different species are provided elsewhere (Rand 1995). Because of their importance to fish populations and as indicators of water quality, aquatic invertebrates have been widely used in the bioassessment of water quality in streams and rivers. Use of aquatic insects is popular for this purpose because they can be observed with the naked eye and are easy to collect. Comparisons of the sensitivities of various aquatic species indicate that aquatic invertebrates are among the most susceptible to the toxic effects of environmental contaminants. The quantification of toxicant stress on aquatic organisms is determined using both acute and chronic tests. Several invertebrate species are consistently used in these tests to represent specific niches in the aquatic environment.

Daphnids (*Daphnia* spp. and *Ceriodaphnia* spp.) are freshwater crustaceans ubiquitous to freshwater environments (lakes, ponds, rivers, and streams). Their importance in aquatic systems is derived from their roles in converting phytoplankton and bacteria into animal protein and serving as a significant food source for numerous fish species. Daphnids are used extensively for acute and chronic toxicity testing for a variety of additional reasons including availability, adaptability to lab conditions, minimal space needs, and sensitivity to toxicants. Their short life cycle facilitates rapid testing and, along with their high fecundity, is an advantage for reproductive toxicity testing.

Amphipods (*Hyalella azteca*), oligochaetes (*Lumbriculus variegatus*), midges (*Chironomus tentans*), clams (*Corbicula fluminea*), and cladocerans are common organisms used in freshwater sediment tests. Endpoints evaluated in these tests include survival, growth, and reproduction. Although the use of some reproductive endpoints can be problematic because of the size of offspring, total numbers of young produced during exposure are often recorded as indicative of direct effects on reproduction as well as adult survival. Midges are perceived to be relatively insensitive to toxicant effects, but data from chronic toxicity tests indicate that midges are equally sensitive as daphnids. Oligochaetes are often used in acute toxicity tests for contaminants in sediments. They are considered relatively tolerant of most contaminant classes. In addition, oligochaetes are used in assessing bioaccumulation potential for contaminants in sediments. Cladocerans are not in direct contact with sediment; however, they are in frequent contact with the sediment surface and thus are exposed to both soluble and sediment-bound contaminants.

A variety of invertebrate species were recently experimentally evaluated and reviewed for the development of water quality criteria for perchlorate (Dean et al. 2004). Those studies are presented in Table 8-1.

Table 8-1 Acute median effective concentrations of perchlorate to various invertebrate animal species

Exposure duration	Common name	Scientific name	Acute median effect concentration ($mg \cdot L^{-1}$ or $mg \cdot kg^{-1}$)
Aquatic			
48-h static	Cladoceran	*Ceriodaphnia dubia*	66 (40–144)[a] 77.8 (not given)[b]
48-h static	Cladoceran	*Daphnia magna*	490 (406–591) [a]
96-h static	Amphipod	*Hyalella azteca*	>1000[c]
96-h flow-through	Oligochaete	*Lumbriculus variegatus*	3710 (3550–3880)[d]
96-h flow-through	Asiatic clam	*Corbicula fluminea*	6680 (5300–8400)[d]
48-h flow-through	Midge	*Chironomus tentans*	8140 (6600–10 000)[d]
Terrestrial			
Acute	Earthworm	*Eisenia fetida*	4450[a]
Subchronic	Earthworm-	*E. fetida*	5071 (4694–5478)[e]

[a] EA Engineering, Science, and Technology 1998

[b] Block Environmental Services 1998

[c] EA Engineering, Science, and Technology 2000

[d] Dean et al. 2004

[e] Parsons 2002

Effects of inorganic contaminants on earthworms

The following studies are a representative selection of investigations that have been performed on the earthworm *Eisenia fetida* using various tests (primarily dermal contact). Currently, only 3 research groups (EA Engineering [EA Engineering, Science, and Technology 1998], Parsons [2002], and Texas Tech University) have conducted tests for effects of perchlorate on *E. fetida*. Most of the studies described below focus on inorganic contaminants. They are being reviewed here because it was hypothesized that some of these test compounds may behave in ways similar to perchlorate.

E. fetida were exposed for 2 d to each of 90 different chemicals using the filter-paper contact test. This organism was used as a marker species to indicate the relative acute toxicities of these 90 chemicals to earthworms and other soil invertebrates (Roberts and Dorough 1984). Chemicals tested included solvents, metals, drugs, and carcinogens, among other compounds. The most relevant to our interest in perchlorate were the compounds that ionize in solution, such as ammonium nitrate, methyl iodide, sodium chloride, and sodium nitrate. Concentrations were expressed in $\mu g \cdot cm^{-2}$, and mortality was the clinical endpoint examined. The acute toxicity of the chemicals tested was categorized to allow the investigators to place a chemical among others of similar acute toxicity without the necessity of recalling specific toxicity values. The system of classification, based on LC50 values, was as follows (Roberts and Dorough 1984):

- supertoxic ($<1.0\ \mu g \cdot cm^{-2}$),
- extremely toxic (1 to 10 $\mu g \cdot cm^{-2}$),
- very toxic (10 to 100 $\mu g \cdot cm^{-2}$),
- moderately toxic (100 ro 1000 $\mu g \cdot cm^{-2}$), and
- relatively nontoxic ($>1000\ \mu g \cdot cm^{-2}$).

Salts of the heavy metals, cadmium, copper, and lead, were very toxic, but high mortality at some lower exposure concentrations indicates that heavy metals border the extremely toxic class of chemicals. Inorganic compounds such as ammonium nitrate, sodium chloride, and sodium nitrate were found to be moderately toxic (100 to 1000 $\mu g \cdot cm^{-2}$). Although ammonium perchlorate and sodium perchlorate were not specifically tested, the acute toxicity of similar compounds such as ammonium nitrate and sodium chloride allows for rough estimation of the acute toxicity categories in which the perchlorate salts would likely be classified.

In a different study, 4 species of earthworm were exposed under 2 test protocols to each of 62 test chemicals (Callahan et al. 1994). *E. fetida* and the 2-d contact test were again used. LC50 values ($mg \cdot kg^{-1}$) for cadmium chloride, cadmium nitrate, copper chloride, copper nitrate, lead nitrate, nickel chloride, nickel nitrate, zinc chloride, and zinc nitrate ranged from 0.563e-3 to 0.632e-2 $mg \cdot kg^{-1}$ (Callahan et al. 1994). This study illustrates that when the cation is a heavy metal, increased lethality is observed. These inorganic chemicals were not entirely similar to ammonium and sodium perchlorate; thus, the LC50 values of these compounds may not provide a good approximation of toxicity of perchlorate.

The effects of cadmium, copper, lead, and zinc on survival, growth, cocoon production, and cocoon viability of *E. fetida* were determined in 2 types of artificial soil experiments, one in which the earthworms were exposed to single metals and one in which the worms were exposed to mixtures of metals (Spurgeon and Hopkin 1995). Review of this investigation is somewhat worthwhile because soils were spiked with solutions of cadmium, copper, lead, or zinc as nitrates, which are inorganic compounds like sodium perchlorate and ammonium perchlorate, although the metal cation is undoubtedly the toxic entity. The investigators determined the concentrations of cadmium, copper, lead, and zinc that increased mortality and reduced growth and cocoon production in the toxicity experiments with *E. fetida*. They determined that cocoon production was the most sensitive parameter, while mortality was the least sensitive (Spurgeon and Hopkin 1995). In addition, it was concluded that zinc was responsible for the majority of effects on earthworms in the artificial soils containing mixtures of metals (Spurgeon and Hopkin 1995).

Acute Earthworm Toxicity Experiments with Perchlorate Salts

Recent studies supported by the US Air Force have tested sodium perchlorate toxicity to *E. fetida* (EA Engineering, Science, and Technology 1998; Parsons 2002). Earthworms were tested in artificial soil in acute (EA Engineering, Science, and Technology 1998) and subchronic (Parsons 2002) scenarios designed to simulate US Environmental Protection Agency (USEPA) Office of Pesticide Program guidelines. Under these conditions, acute and subchronic LC50s for sodium perchlorate to *E. fetida* were 4450 $mg \cdot kg^{-1}$ and 5071 $mg \cdot kg^{-1}$, respectively.

A variety of earthworm tests were conducted on perchlorate salts through Strategic Environmental Research and Development Program (SERDP)-funded studies at Texas Tech University. A summary of the results of the studies is presented below and is also available in a recent publication (Landrum et al. 2006).

Filter-paper contact tests

Adult *E. fetida*, those with distinguishable clitella, were exposed to varying concentrations of sodium or ammonium perchlorate. The earthworms were depurated for 24 h, then exposed to the test substances or to 1 mL Milli-Q water (controls) in flat-bottom, 60-mL Opticlear glass vials as described by Organisation for Economic Co-operation and Development guidelines (OECD 1984). Earthworms were placed individually in vials because the death of 1 worm might have adverse effects on others in the same vial (OECD 1984). Mortality, the clinical endpoint, was recorded daily during each 14-d study.

In the initial range-finding experiment, sodium perchlorate was lethal to all replicates above 7000 mg·L^{-1} (Figure 8-1). Forty-five percent of total mortalities occurred within 24 h. All mortalities that occurred in this time period were replicates exposed to the highest concentrations. Mortality in the other treatment groups was spread over the following days. Based on these data, the acute LC50 (calculated via probit analysis) for sodium perchlorate to *E. fetida* was 4150 mg·L^{-1}.

In the initial range finding experiment with ammonium perchlorate, the test chemical was lethal to all replicates above 1000 mg·L^{-1} (Figure 8-2). From these data, an acute LC50 of 4140 mg·L^{-1} ammonium perchlorate was calculated (probit analysis) for *E. fetida*. Approximately 58% of total mortalities occurred within 24 h with complete mortality in the 50 000 and 100 000 mg·L^{-1} treatment groups. Mortalities in the 10 000 mg·L^{-1} treatment group occurred between experimental days 7 and 11. Mortality

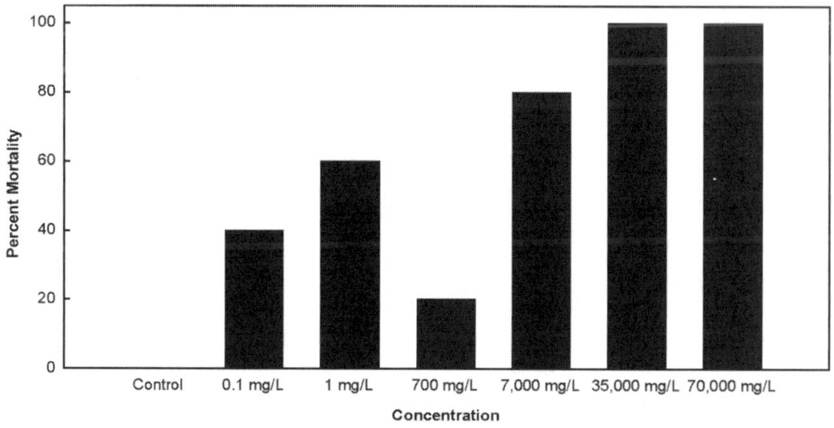

Figure 8-1 Earthworm (*Eisenia fetida*) percent mortality at concentrations following a 14-d filter-paper contact test using sodium perchlorate

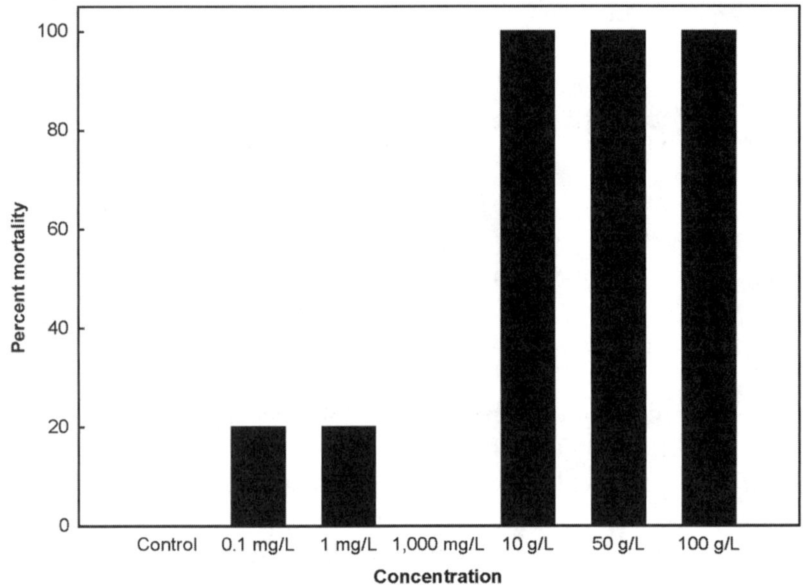

Figure 8-2 Earthworm (*E. fetida*) percent mortality at various concentrations following a 14-d filter-paper contact test using ammonium perchlorate

in the remaining treatment groups ($0.1\ mg \cdot L^{-1}$ and $1\ mg \cdot L^{-1}$) occurred on day 11.

In the initial range-finding experiment with sodium perchlorate, there was a significant difference ($T = 4.57$, $p = 0.0006$, df = 12) between average earthworm weight before and after exposure. This difference can be attributed to weight loss that occurred in the worms that died in the treatment groups exposed to the highest concentrations. Worms that died throughout the experiment lost considerably more weight than those that survived the 14 days. Thus, there were significant differences ($p = 0.0116$, $p = 0.0001$, and $p < 0.0001$) in pre- and post-exposure mean weights for worms exposed to the greatest concentrations (7, 35, and $70\ g \cdot L^{-1}$, respectively), in which mortality also was high (80%, 100%, and 100%, respectively) (Figure 8-3).

In the initial range-finding experiment for ammonium perchlorate, there was a significant difference ($T = 3.48$, $p = 0.0045$, df = 12) between average earthworm weight before and after exposure. The majority of this difference can be attributed to weight loss that occurred in worms that died in the treatment groups exposed to the greatest concentrations. There were significant differences in pre- and post-exposure mean weights

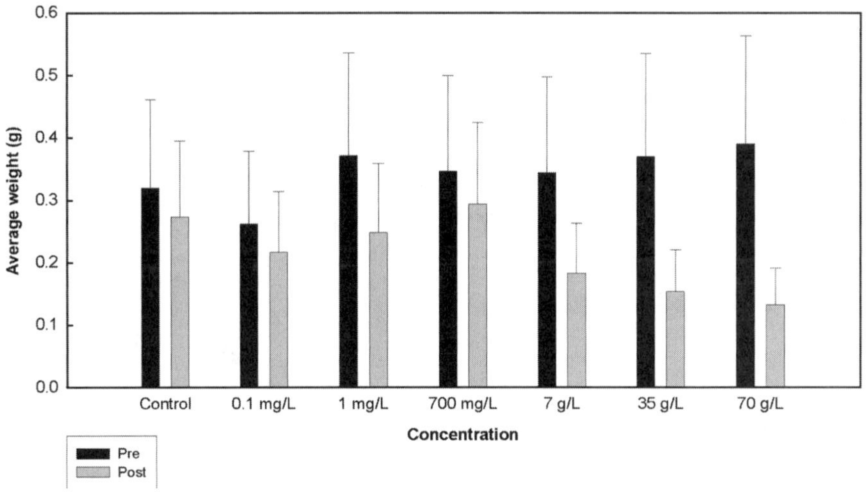

Figure 8-3 Average weight (g) of *E. fetida* before and after 14-d exposure to sodium perchlorate-treated filter paper. Post averages include weights from both surviving earthworms and those that died throughout the test. Error bars represent 1 standard deviation of the mean.

($p < 0.0001$) for worms exposed to the greatest concentrations (10, 50, and 100 $g \cdot L^{-1}$) in which mortality also was high (100%) (Figure 8-4).

Sand tests

Adult *E. fetida* were exposed to various concentrations of sodium perchlorate in sand. For each treatment, consisting of 1 worm per vial, 10 replicates were constructed. Control vials (also 10 replicates) were treated with 1 mL of Milli-Q water. Worms were depurated for 24 h before being placed in test vials and were washed, dried, and weighed before use. Mortality was the clinical endpoint examined during the 14-d experiment.

Mortality was consistent at 0% in the lowest concentration treatment groups, and sodium perchlorate was determined to be lethal to all replicates exposed to a concentration greater than 2000 $mg \cdot kg^{-1}$ (Figure 8-5). From these data, an acute LC50 of 2550 $mg \cdot kg^{-1}$ sodium perchlorate was calculated (probit analysis) for *E. fetida*. Approximately 67% of total mortalities occurred within 24 h. All of the mortalities that occurred in this time period were replicates exposed to the highest concentrations. Ninety percent of the 10 000 $mg \cdot kg^{-1}$ exposure replicates and 70% of the 5000 $mg \cdot kg^{-1}$ exposure replicates died within the first 24 h. The remaining replicates of those treatment groups died in the next 2 days. Mortalities in

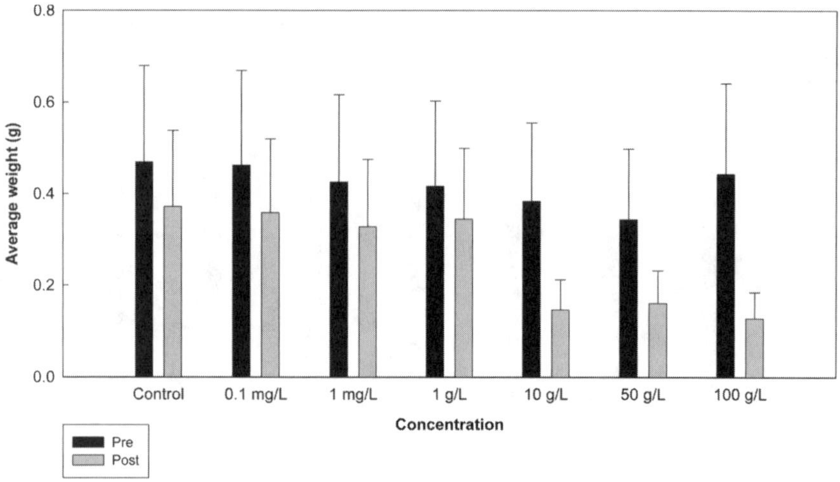

Figure 8-4 Average weight (g) of *E. fetida* before and after 14-d exposure to ammonium perchlorate-treated filter paper. Post-exposure averages include weights from both surviving earthworms and those that died throughout the test. Error bars represent 1 standard deviation of the mean.

the next highest concentration treatment group (2000 $mg \cdot kg^{-1}$ exposure) occurred on experimental days 6, 8, and 12. The sole mortality in the 1000 $mg \cdot kg^{-1}$ treatment group also occurred on day 12.

There was no significant difference ($T = 2.82$, $p = 0.0155$, df = 12) in average earthworm weight before and after exposure to sodium perchlorate. Although there was a significant difference ($p = < 0.0001$ and $p = 0.0009$) between mean earthworm weight before and after exposure to 5000 $mg \cdot kg^{-1}$ and 10 000 $mg \cdot kg^{-1}$ sodium perchlorate, the weight loss in worms that died in these treatment groups did not have as great an impact on overall average earthworm weight as did the weight loss in earlier experiments involving sodium perchlorate (Figure 8-6).

Although it was not expected, cocoon production was observed in this experiment in the lowest concentration treatment groups, in which there also were no mortalities (Figure 8-7). Because vials contained only 1 worm, the earthworms must have copulated prior to being separated and exposed. As previously stated, earthworm copulation does not directly lead to fertilization (Gaddie and Douglas 1975), so the earthworms probably copulated before being collected from the colony or during depuration in the holding container. The earthworms were then randomly placed in their individual vials, where the egg cells and sperm matured and the cocoon-laying process ensued. Average cocoon production was highest in the control group (average = 1.4) and almost identical in the 100 $ng \cdot g^{-1}$

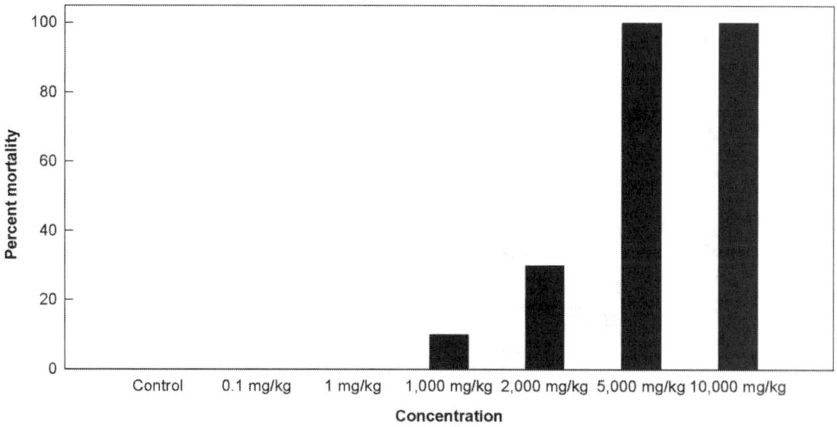

Figure 8-5 Earthworm (*E. fetida*) percent mortality at various concentrations following a 14-d spiked sand test using sodium perchlorate

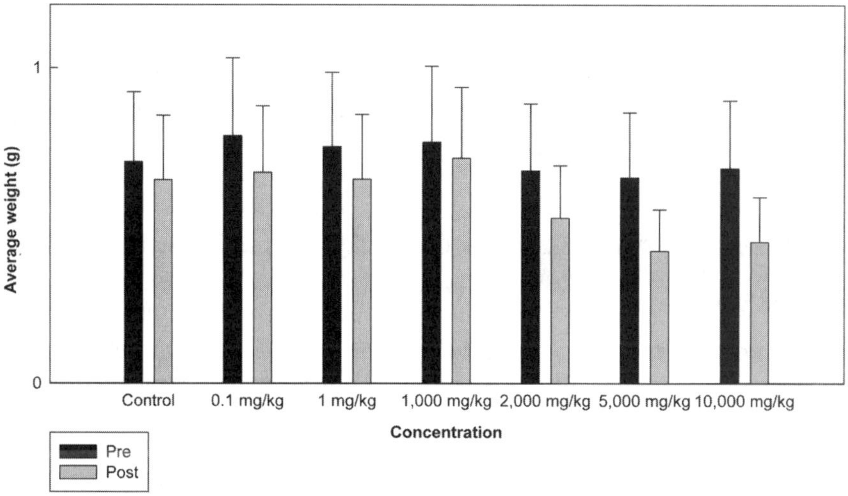

Figure 8-6 Average weight (g) of *E. fetida* before and after 14-d exposure to sodium perchlorate-treated sand. Post-exposure averages include weights from both surviving earthworms and those that died throughout the test. Error bars represent 1 standard deviation of the mean.

(average = 0.6) and 1 $mg \cdot kg^{-1}$ exposure groups (average = 0.8). There was no significant difference between the control mean and the 1 $mg \cdot kg^{-1}$ mean. However, there was a significant difference between the control mean and the 100 $ng \cdot g^{-1}$ mean ($p > 0.05$). From these data, an acute EC50 was calculated (via probit analysis) at 260 $mg \cdot kg^{-1}$ sodium perchlorate for *E. fetida* reproduction (cocoon production). Of the treatment

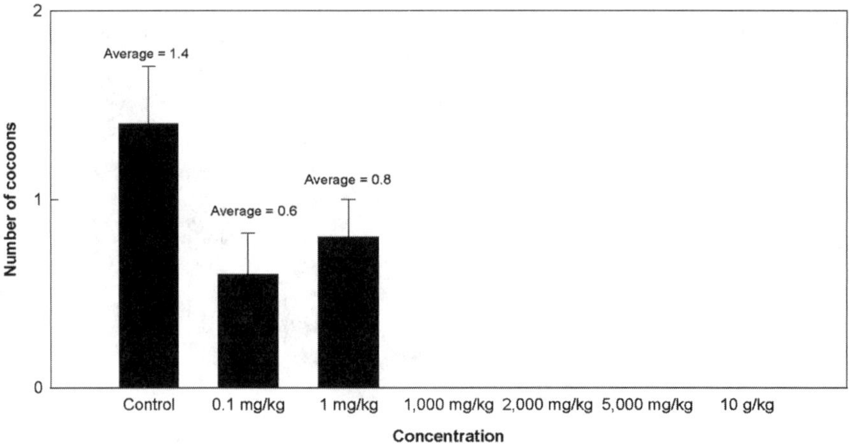

Figure 8-7 Average number of earthworm cocoons produced per jar in a 14-d test in which earthworms were exposed to sodium perchlorate. Error bars represent 1 standard deviation of the mean

groups in which cocoons were produced, hatchlings emerged in both the control and the 1 $mg \cdot kg^{-1}$ exposure groups. Hatchling emergence was highest in the control group.

Earthworm Reproductive Toxicity Experiments

Artificial soil tests

Using a modified test procedure developed by van Gestel et al. (1989), the effect of sodium perchlorate on reproduction of *E. fetida* was examined. The test involved exposing earthworms to artificial soil dosed with sodium perchlorate to evaluate whether perchlorate would affect cocoon production. Before exposure, earthworms were acclimated for 1 week in untreated artificial soil. Earthworms were then exposed for 3 to 4 weeks to the test substance incorporated in the artificial soil. At the end of each 3- or 4-week testing period, the number of cocoons produced, the number of juvenile worms, the number of alive and dead adult worms, and worm weights were determined. To determine the influence of the chemical substance on cocoon hatchability, cocoons were incubated for 5 weeks in untreated artificial soil. At the conclusion of the 5-week period, the soil was searched and the number of cocoons and juveniles determined.

In the initial experiment involving sodium perchlorate, there was no significant difference ($T = 2.25$, $p = 0.0546$, df = 8) between pre- and post-exposure weights for all of the concentrations tested (Figure 8-8). Only the 10 mg·kg^{-1} treatment group demonstrated a significant difference in mean weights within a treatmen group before and after exposure (p = 0.0013).

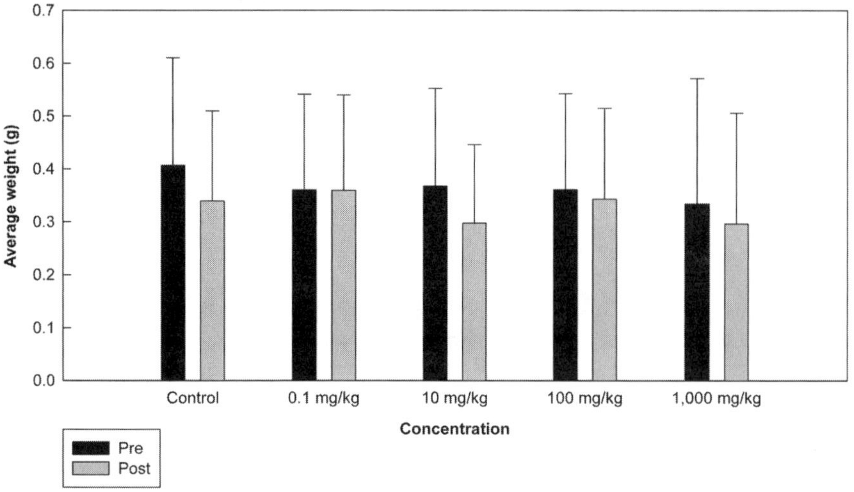

Figure 8-8 Average weight (g) of *E. fetida* before and after 3-week exposure to sodium perchlorate-treated artificial soil. Post-exposure averages include weights from both surviving earthworms and those that died throughout the test. Error bars represent 1 standard deviation of the mean.

Likewise, no significant difference (T = 1.88, p = 0.0964, df = 8) between pre- and post-exposure weights was observed in a second artificial soil experiment (Figure 8-9). Although significant differences in mean weights were observed in the control and 100 ng·g^{-1} exposure groups (p = 0.0007 and p = 0.0011), reductions in mean weights in those concentration exposures were not significant overall.

Conversely, in a third artificial soil experiment, a significant difference (T = 4.59, p = 0.0018, df = 8) was observed between pre- and post-exposure mean weights (Figure 8-10). The difference was due to the significant reductions in post-exposure weights observed in the control, 100 ng·g^{-1}, 10 mg·kg^{-1}, and 100 mg·kg^{-1} exposure groups (p = 0.0150, p = 0.0093, p = 0.0001, and p = 0.0033).

The highest average cocoon production was observed in the control group and the 10 mg·kg^{-1} exposure group (average number of cocoons per treatment group = 2.0). Cocoon production was less in the 0.1 mg·kg^{-1}

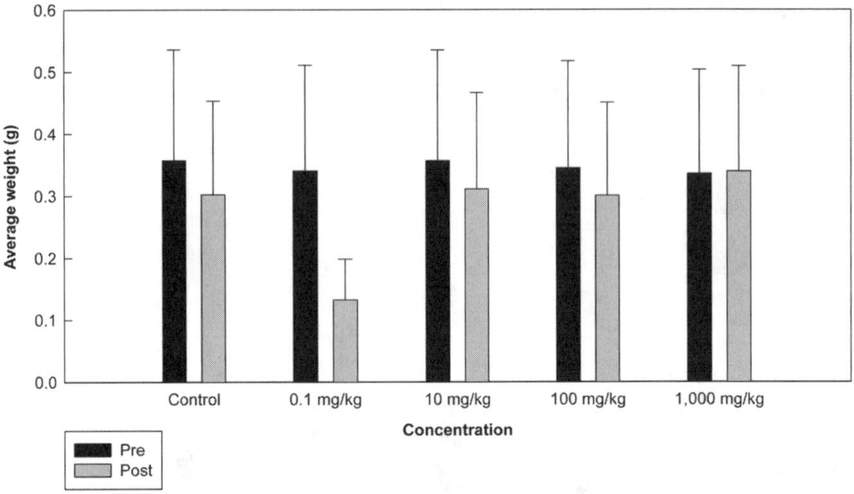

Figure 8-9 Average weight (g) of *E. fetida* before and after 3-week exposure to sodium perchlorate-treated artificial soil. Post-exposure averages include weights from both surviving earthworms and those that died throughout the test. Error bars represent 1 standard deviation of the mean.

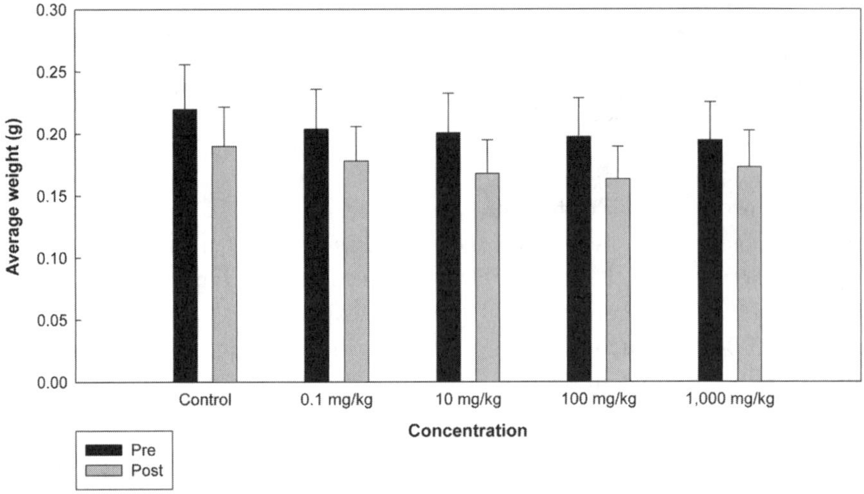

Figure 8-10 Average weight (g) of *E. fetida* before and after 4-week exposure to sodium perchlorate-treated artificial soil. Post-exposure averages include weights from both surviving earthworms and those that died throughout the test. Error bars represent 1 standard deviation of the mean.

(average = 1.25) and 100 $mg \cdot kg^{-1}$ (average = 0.75) treatment groups. No cocoon production was observed in the 1000 $mg \cdot kg^{-1}$ treatment group. There were no significant differences among treatment group means in which cocoons were produced. From these data, a subchronic (28-d) EC50 of 350 mg·kg^{-1} sodium perchlorate was calculated (probit analysis) for *E. fetida* reproduction (cocoon production). Of the treatment groups in which cocoons were produced, hatchlings emerged in both the control and the 10 mg·kg^{-1} exposure groups. Hatchling emergence was highest in the control group.

Sand tests

Adult *E. fetida* were exposed to various concentrations of sodium perchlorate in layered media consisting of sand and a small amount of manure. Sodium perchlorate was determined to be lethal to all replicates exposed to concentrations greater than 1000 mg·kg^{-1} (Figure 8-11). From these data, a subchronic (21-d) LC50 of 500 mg·kg^{-1} sodium perchlorate calculated (probit analysis) for *E. fetida*. Approximately 36% of total mortalities occurred within 24 h. All of the mortalities that occurred in this time period were in treatment groups exposed to the highest concentrations. Ninety percent of the worms exposed to 10 000 mg·kg^{-1} and 70% of the worms exposed to 5000 mg·kg^{-1} sodium perchlorate died within the first 24 h. Mortalities in the other treatment groups were spread out over the following days.

Significant reductions in earthworm weights after exposure to sodium perchlorate were observed in all treatment groups in the sand trial (Figure 8-12). Overall, there was a significant difference ($T = 7.05$, $p = < 0.0001$, df = 12) between worm weights before and after the exposure period.

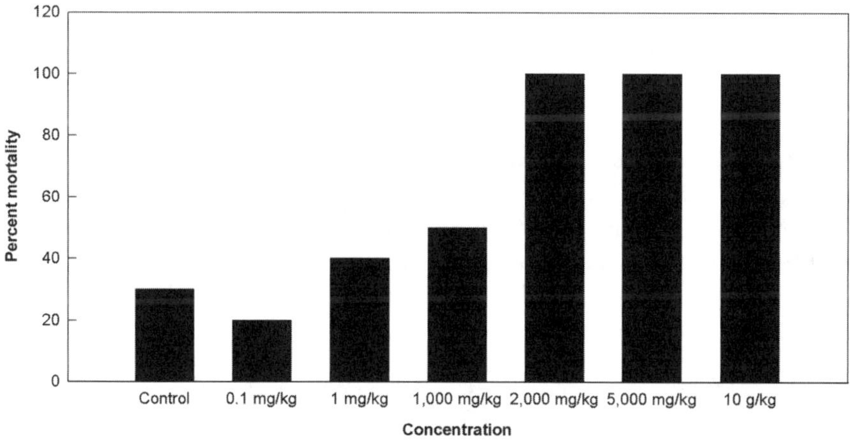

Figure 8-11 Earthworm (*E. fetida*) percent mortality at various concentrations following a 3-week spiked sand test using sodium perchlorate

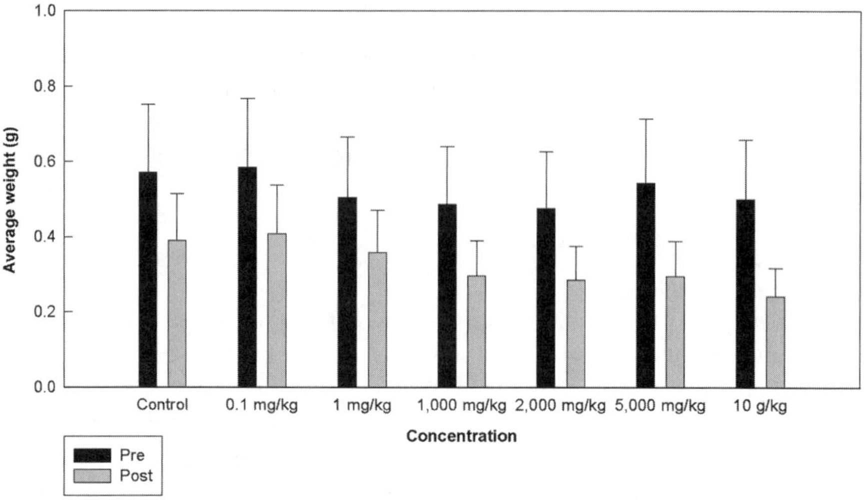

Figure 8-12 Average weight (g) of *E. fetida* before and after 3-week exposure to sodium perchlorate-treated sand. Post-exposure averages include weights from both surviving earthworms and those that died throughout the test. Error bars represent 1 standard deviation of the mean.

Cocoon production was also observed in the control and the 0.1 $mg \cdot kg^{-1}$ exposure groups. Average cocoon production was higher in the control group (average = 0.50) than in the 0.1 $mg \cdot kg^{-1}$ treatment group (average = 0.20). There were no significant differences between the means for the treatment groups. From these data, a subchronic (21-d) EC50 was calculated (probit analysis) at 1.3 $mg \cdot kg^{-1}$ sodium perchlorate for *E. fetida* reproduction (cocoon production). These cocoons were not incubated to assess hatchability.

Discussion

The effects of the perchlorate anion on *E. fetida* survival and reproductive success were evaluated with tests using 3 different exposure media: filter paper, artificial soil, and sand (Table 8-2). All studies utilized a range of perchlorate concentrations in order to simulate levels that are likely to occur in the environment under different scenarios (typical soil levels in contaminated areas versus spill levels).

The OECD filter-paper contact test involved exposing earthworms to the test compound on moist filter paper for 14 d in order to evaluate earthworm survival. A similar test involved exposing the earthworms to perchlorate-spiked sand. Two filter-paper dermal tests used to evaluate the toxicity of sodium perchlorate and ammonium perchlorate to *E. fetida*,

Table 8-2 Summary of acute and reproductive earthworm toxicity experiments conducted with perchlorate salts at Texas Tech University

Test type (exposure duration)	Test chemical	LC50 or EC50	Other observations
Filter-paper contact			
Acute	Sodium perchlorate	4150 $mg \cdot L^{-1}$	Reduction in earthworm weight at concentrations >700 $mg \cdot L^{-1}$
Acute	Ammonium perchlorate	4140 $mg \cdot L^{-1}$	Reduction in earthworm weight at concentrations >0.1 $mg \cdot L^{-1}$
Acute	Sodium perchlorate	NA	Reduction in earthworm weight at concentrations >1000 $mg \cdot L^{-1}$
Acute	Ammonium perchlorate	NA	Reduction in earthworm weight at concentrations >5000 $mg \cdot L^{-1}$
Sand			
Acute	Sodium perchlorate	2550 $mg \cdot kg^{-1}$	Reduction in earthworm weight at concentrations >2000 $mg \cdot kg^{-1}$
Subchronic	Sodium perchlorate	500 $mg \cdot kg^{-1}$	
Reproductive tox			
Acute, sand	Sodium perchlorate	260 $mg \cdot kg^{-1}$	No weight reduction at the concentrations tested (up to 1000 $mg \cdot kg^{-1}$)
Subchronic, soil	Sodium perchlorate	350 $mg \cdot kg^{-1}$	No cocoon production at the concentrations tested >100 $mg \cdot kg^{-1}$
Subchronic, sand	Sodium perchlorate	1.3 $mg \cdot kg^{-1}$	No cocoon production at the concentrations tested >1 $mg \cdot kg^{-1}$

were inconclusive according to the validity criterion for mortality testing in the OECD guidelines, which states that mortality among the controls should not exceed 10% (OECD 1984). However, in both tests that failed this criterion, survival in the control group was the highest when compared to that in the other treatment groups. Data from a second set of filter-paper contact tests met the validity criteria and were used to estimate an acute LC50 of 4150 $mg \cdot L^{-1}$ for sodium perchlorate and 4140 $mg \cdot L^{-1}$ for ammonium perchlorate.

In the tests that met the OECD test criterion for mortality in the controls, earthworm survival decreased as concentration of perchlorate increased, with no worms surviving 14 d at the highest treatment concentrations (>2000 $mg \cdot L^{-1}$). A notable difference between the tests is the percent survival in the 0.1 $mg \cdot L^{-1}$ and 1 $mg \cdot L^{-1}$ concentration groups. It should be emphasized that the number of replicates per concentration was different between these tests (5 versus 10). Thus, the deaths of a few individuals had a more marked impact in the former as compared to the latter trial. Experiments that use the greatest number of replicates are suggested because replication allows uncontrollable unit-to-unit variation (natural variation) to be estimated in an experiment, which can then be used as the basis for comparison of treatment effects (Sparks 2000).

The concentrations determined to affect survival of *E. fetida* (>2000 $mg \cdot L^{-1}$) are likely to occur only under extreme conditions. Mortality at concentrations of this magnitude is more likely attributable to the physical stress on the earthworms from salt intolerances rather than to the toxicity of perchlorate itself. Although these concentrations might seem excessive, perchlorate levels have approached these concentrations in some soil samples analyzed. For instance, perchlorate-contaminated soil sampled from a former propellant-mixing facility at the Naval Weapons Industrial Reserve Plant (NWIRP) in McGregor, Texas ranged from 0.023 to 1800 ppm ($mg \cdot kg^{-1}$) (Motzer 2001), and similar perchlorate levels were measured in soils from the Las Vegas Wash (Smith et al. 2004).

In a comparable study, *E. fetida* were exposed in a 2-d filter-paper contact test to compounds that would ionize in solution, similar to sodium perchlorate (Roberts and Dorough 1984). These compounds were methyl iodide, sodium chloride, ammonium nitrate, and sodium nitrate. LC50 values for these compounds, based on a system of classification designed to allow the investigators to place a chemical among others of similar toxicity, indicate that they are moderately toxic (100 to 1000 $\mu g \cdot cm^{-2}$). Based on this system of classification, sodium perchlorate would be classified as relatively nontoxic (>1000 $\mu g \cdot cm^{-2}$). When compared to data collected by Roberts and Dorough (1984), ammonium perchlorate would be classified as relatively nontoxic (>1000 $\mu g \cdot cm^{-2}$), whereas ammonium nitrate would be somewhat more toxic (moderately toxic: 100 to 1000 $\mu g \cdot cm^{-2}$).

In another investigation using the 2-d contact test, *E. fetida* were exposed to cadmium chloride, cadmium nitrate, copper chloride, copper nitrate, lead nitrate, nickel chloride, nickel nitrate, zinc chloride, and zinc nitrate (Callahan et al. 1994). The LC50 values calculated for these compounds ranged between 0.56 to 0.63 $\mu g \cdot kg^{-1}$ (ppb) and are significantly lower than the LC50 value calculated for methyl iodide, sodium chloride, sodium nitrate, and both sodium and ammonium perchlorate. Therefore, smaller amounts of these metal salts are necessary to cause a benchmark degree of mortality compared to other compounds tested.

Perchlorate had some effect on weight retention when average weight before and after exposure were compared. In the filter-paper contact and sand tests, significant differences between pre-exposure and post-exposure weights were observed when *E. fetida* lost weight and died in some or all of the treatment groups, resulting in significantly reduced post-exposure weights.

The effect of perchlorate on the reproductive success (cocoon production) of *E. fetida* over a 3- to 4-week test period was also examined. Earthworms will increase their number if force-bred by placing a large number of mature individuals in a given area (i.e., 1-L jar). Propagation occurs because overpopulation of a given area enables earthworms to be closer to each other so that they do not have to search for a mate (Gaddie and Douglas 1975). Thus, earthworms in these experiments were provided with numerous opportunities to propagate because they were in close contact with other individuals in 1-L jars both during the 1-week acclimation period in untreated soil and throughout the 3- to 4-week exposure to sodium perchlorate.

In the artificial soil experiment, production of cocoons was observed in soil containing up to 100 $mg \cdot kg^{-1}$ perchlorate, with no production in the uppermost treatment group (1000 $mg \cdot kg^{-1}$). Cocoon production was highest in the control group and the 10 $mg \cdot kg^{-1}$ treatment groups, although overall cocoon production was low. From these data, an EC50 estimate of 350 $mg \cdot kg^{-1}$ was calculated. In comparison, van Gestel et al. (1989) exposed *E. fetida* to copper chloride in a standardized artificial soil experiment in which the worms, after being preconditioned for 1 week in untreated soil, were exposed to the chemical for 3 weeks. The test was performed at 5 test concentrations (60, 120, 180, 240, and 300 $mg \cdot kg^{-1}$) and a control. The investigators also determined an EC50 value for the effect of copper chloride following an additional preconditioning period in treated artificial soil and following the 3-week exposure. These EC50 values were 62 (21–181) and 191 (163–223) $mg \cdot kg^{-1}$ dry artificial soil, for the worms preconditioned for 1 week or for a longer period, respectively (van Gestel et al. 1989).

The cocoons from the final artificial soil experiment were incubated in clean soil for 5 weeks to evaluate cocoon viability and successful emergence of juveniles from the cocoons. Unlike the study by van Gestel et al. (1989) in which copper chloride did not affect cocoon viability at any of the concentrations tested, juveniles emerged in both the control and the 10 $mg \cdot kg^{-1}$ exposure concentrations, with the highest emergence in the controls. Interestingly, this pattern paralleled that observed for the unexpected cocoon production in an earlier experiment in which worms were exposed to sodium perchlorate in sand to assess survival. In this experiment, cocoon production was observed up to 1 $mg \cdot kg^{-1}$ perchlorate, with no production in the uppermost treatment groups (≥1000 $mg \cdot kg^{-1}$). From the data collected, an EC50 value of 260 $mg \cdot kg^{-1}$ was estimated. After incubation in clean sand for a little over 5 weeks, juveniles emerged in both control and 1 $mg \cdot kg^{-1}$ concentrations, with the highest emergence observed in the controls. In both the final artificial soil test and the acute toxicity sand test there was no juvenile emergence from cocoons produced in 100 $ng \cdot g^{-1}$ perchlorate.

The last cocoon production test mirrored the above experiment in which worms were exposed to sodium perchlorate in sand to assess survival. Cocoon production in the final earthworm reproduction experiment was observed up to 0.1 $mg \cdot kg^{-1}$ perchlorate, with no production in the uppermost treatment groups (≥1 $mg \cdot kg^{-1}$). Although these results were similar to those observed in earlier tests, cocoon production was somewhat lower. From these data, an EC50 value of 1.3 $mg \cdot kg^{-1}$ was estimated. There was insufficient time to incubate these cocoons in untreated sand to assess cocoon viability.

In addition to the reproductive toxicity data collected, opportunistic acute toxicity data were gathered. Sodium perchlorate was determined to be lethal to all replicates greater than 1000 $mg \cdot kg^{-1}$. From these data, a subchronic LC50 value of 500 $mg \cdot kg^{-1}$ was calculated.

In the final artificial soil experiment, a significant difference was observed between pre- and post-exposure mean weights. The difference was largely due to the significant reductions in post-exposure weights observed in the control, 0.1 $mg \cdot kg^{-1}$, 10 $mg \cdot kg^{-1}$, and 100 $mg \cdot kg^{-1}$ exposure groups ($p = 0.0150$, $p = 0.0093$, $p = 0.0001$, and $p = 0.0033$). Likewise, in the sand jar and vial trials, significant differences between pre- and post-exposure weights were observed as a result of significant reductions in earthworm weights after exposure to sodium perchlorate in some if not all treatment groups.

Conclusions and Research Recommendations

Studies conducted previously (EA Engineering, Science, and Technology 1998, 2000) suggested that perchlorate (as sodium perchlorate) is not particularly lethal to earthworms (*E. fetida*), with acute (14-d) and subchronic (28-d) LC50s of 4450 $mg \cdot kg^{-1}$ and 5071 $mg \cdot kg^{-1}$, respectively. Our observations are similar to those reported previously. The acute (14-d dermal contact test) LC50s for sodium perchlorate and ammonium perchlorate to *E. fetida* were 4150 $mg \cdot kg^{-1}$ and 4140 $mg \cdot kg^{-1}$, respectively. More realistic exposure scenarios (14-d sand tests) and slightly longer exposure times (21 d) produced data suggesting that perchlorate was slightly more toxic under these conditions; the acute (14-d sand test) LC50 for sodium perchlorate to *E. fetida* was 2550 $mg \cdot kg^{-1}$ and the subchronic (21-d) LC50 for sodium perchlorate to *E. fetida* was 500 $mg \cdot kg^{-1}$.

Results of reproductive toxicity tests were more variable but indicated that cocoon production was more sensitive to perchlorate exposure. The acute (14-d sand tests) EC50 for sodium perchlorate to *E. fetida* reproduction (cocoon production) was 260 $mg \cdot kg^{-1}$. In the treatments that reproduced, cocoon production was slightly higher at 1 $mg \cdot kg^{-1}$ than at 1 $ng \cdot g^{-1}$. Highest cocoon production was observed in the controls. The sub-chronic (21-d sand tests) EC50 for sodium perchlorate to *E. fetida* reproduction (cocoon production) was 1.3 $mg \cdot kg^{-1}$. The subchronic (28-d artificial soil tests) EC50 for sodium perchlorate to *E. fetida* reproduction (cocoon production) was 350 $mg \cdot kg^{-1}$. In the treatments that reproduced, cocoon production was slightly higher in the control and 10 $mg \cdot kg^{-1}$ treatments than in the 0.1 $mg \cdot kg^{-1}$ and 100 $mg \cdot kg^{-1}$ treatments.

The number of studies evaluating perchlorate effects on aquatic and terrestrial invertebrates has increased during the last couple of years. However, the database on invertebrates is still relatively small. The data accumulated to date suggest that perchlorate is not lethal to earthworms at environmentally relevant soil concentrations (exceptions being spills or point sources). Perchlorate does appear to affect earthworm reproduction at concentrations that would have environmental relevance. In addition, chronic perchlorate exposure may reduce earthworm weight, which could influence survival over the long term. It is recommended that endpoints other than lethality be examined in order to determine the subtle effects that perchlorate may have on terrestrial invertebrates.

References

Abdul Rida AMM, Bouche MB. 1995. Earthworm contribution to ecotoxicological assessments. Acta Zool Fennica 196:307–310.

Barron MG. 2003. Bioaccumulation and bioconcentration in aquatic organisms. In Hoffman DJ, Rattner BA, Burton GA, Cairns J, editors. Handbook of ecotoxicology. Washington (DC): Lewis. p 877–892.

Benke AC. 1996. Secondary production of macroinvertebrates. In: Hauer FR, Lamberti GA, editors. Methods in stream ecology. New York (NY): Academic Press. p 557–578.

Block Environmental Services. 1998. LC50 aquatic toxicity test results for ammonium perchlorate: a two-species chronic definitive bioassay. Final report. San Jose (CA): Santa Clara Valley Water District.

Callahan CA, Shirazi MA, Neuhauser EF. 1994. Comparative toxicity of chemicals to earthworms. Environ Toxicol Chem 13:291–298.

Dean KE, Palacheck RM, Noel JM, Warbritton R, Aufderheide J, Wireman J. 2004. Development of freshwater water-quality criteria for perchlorate. Environ Toxicol Chem 23:1441–1451.

EA Engineering, Science, and Technology. 1998. Results of acute and chronic toxicity testing with sodium perchlorate. Report 2900. Brooks Air Force Base, TX, USA.

EA Engineering, Science, and Technology. 2000. Results of chronic toxicity testing with sodium perchlorate using *Hyalella azteca* and *Pimephales promelas*. Report 3505. Brooks Air Force Base, TX, USA.

Edwards CA. 1983. Development of a standardized laboratory method for assessing the toxicity of chemical substances to earthworms. Brussels, Belgium: Commission of the European Communities.

Edwards CA, Lofty JR. 1977. Biology of earthworms, 2nd ed. New York (NY): J Wiley.

Gaddie RES, Douglas DE. 1975. Earthworms for ecology and profit. Ontario (CA): Bookworm.

Hauer FR, Resh VH. 1996. Benthic macroinvertebrates. In: Hauer FR, Lamberti GA, editors. Methods in stream ecology. New York (NY): Academic Press. p 339–369.

Kendall RJ, Dickerson RL, Giesy JP, Suk WP. 1998. Principles and processes for evaluating endocrine disruption in wildlife. Pensacola (FL): Society of Environmental Toxicology and Chemistry (SETAC).

Landrum M, Canas JE, Coimbatore G, Cobb GP, Jackson WA, Zhang B, Anderson TA. 2006. Effects of perchlorate on earthworm (*Eisenia fetida*) survival and reproductive success. Sci Tot Environ 363:237–244.

Motzer WE. 2001. Perchlorate: problems, detection, and solutions. Environ Forensic 2:301–311.

Neuhauser EF, Lower C, Malacca MR, Milligan DL, Durkin PR. 1985. *Eisenia fetida*. J Environ Qual 14:383–388.

[OECD] Organisation for Economic Co-operation and Development. 1984. OECD guidelines for testing chemicals, No. 207, earthworm acute toxicity test. Paris (FR): OECD.

Paoletti MG. 1999. The role of earthworms for assessment of sustainability and as bioindicators. Agricul Ecosyst Environ 74:137–155.

Parsons. 2002. Sub-chronic toxicity and bioconcentration of perchlorate by the earthworm *Eisenia fetida*. Contract F41624-01-D-8009, Brooks Air Force Base, TX.

Rand GM. 1995. Fundamentals of aquatic toxicology: effects, environmetal fate, and risk assessment. Washington (DC): Taylor & Francis.

Roberts BL, Dorough HW. 1984. Relative toxicities of chemicals to the earthworm *Eisenia fetida*. Environ Toxicol Chem 3:67–78.

Smith PN, Theodorakis CW, Anderson TA, Kendall RJ. 2001. Preliminary assessment of perchlorate in ecological receptors at the Longhorn Army Ammunition Plant (LHAAP), Karnack, Texas. Ecotoxicology 10:305–313.

Smith PN, Yu L, McMurry ST, Anderson TA. 2004. Perchlorate in water, soil, vegetation, and rodents collected from the Las Vegas Wash, Nevada, USA. Environ Pollut 132:121–127.

Sparks T. 2000. Statistics in ecotoxicology. New York (NY): J Wiley.

Spurgeon DJ, Hopkin SP. 1995. Extrapolation of the laboratory-based OECD earthworm toxicity test to metal-contaminated field studies. Ecotoxicology 4:190–205.

Tan K, Anderson TA, Jackson WA. 2004. Degradation kinetics of perchlorate in sediments and soils. Water Air Soil Pollut 151:245–259.

van Gestel CAM, van Dis WA, van Breemen EM, Sparenburg PM. 1989. Development of a standardized reproduction toxicity test with the earthworm species *Eisenia fetida* andrei using copper, pentachlorophenol, and 2,4-dichloroaniline. Ecotoxicol Environ Saf 18:305–312.

nine

Application of Population Models to Perchlorate Ecotoxicology

Stephen B Cox, Kenneth R Dixon, Ronald J Kendall

Introduction

There has been increasing emphasis on the need for consideration of population-level effects in environmental toxicology and risk assessment (Caswell 1996; Forbes and Calow 1999; Pastorok and Ginzburg 2002). This emphasis is partly because the effects of contaminants on populations generally are considered more important than the effects on individuals (Suter II and Barnthouse 2000). Although responses to toxics occur at the individual level, population responses integrate all of the lower-level (organismal, physiological, molecular, etc.) responses and provide an ecologically relevant and consistent framework for interpreting and comparing toxicological information. For example, behavioral, developmental, reproductive, and metabolic effects all potentially contribute to effects on populations. Population-level effects also provide a more tangible measure of effect, especially when regulatory decision processes are being dealt with. If we assume that perchlorate reduces circulating T_4 levels, what does this mean in terms of the ability of animals to forage, reproduce, and in general, carry out the normal activities of life? In contrast, if we know that perchlorate enhances the probability of population extinction for an endangered species, we have a more concrete, and relevant, measure of impact.

Although there is some disagreement about the most appropriate definition of a population, we will simply define a population to be a group of individuals of the same species living together in the same place at the same time. Perchlorate, or any other contaminant, may have both direct and in-

Perchlorate Ecotoxicology. Ronald J Kendall and Philip N Smith, editors.

ISBN-10: 1-880611-90-2 ISBN-13: 978-1-880611-90-6

direct effects on populations (Figure 9-1). In order to have direct effects, perchlorate must alter survival, reproduction, or growth. However, perchlorate may also alter the biotic environment in which populations occur. For example, it may alter individuals' ability to compete for resources, avoid predators, or resist infections. Alternatively, perchlorate may alter the abiotic environment via effects on habitat quality. As a result, the spatial distribution of the population may be altered. These indirect effects are transparent when toxicological information is based solely on individual-level data.

Currently, little to no work has been done on the population-level effects of perchlorate, resulting partly from the relatively recent arrival of perchlorate as a contaminant of concern and partly from the scant evidence of direct effects of perchlorate on aspects of survival and reproduction. Nevertheless, latent effects of chronic exposure to perchlorate, as well as the secondary effects of perchlorate via effects on thyroid function, may have important implications for natural populations. Moreover, the tools for examining population effects are being developed. For example, a variety of modeling frameworks have been proposed to study the impacts of contaminants on populations of wildlife species (Kendall et al. 1994).

Long-term monitoring of natural populations would provide insight into the impacts of perchlorate on natural populations. However, several factors limit the ability of such monitoring studies to enhance understanding of possible impacts. First, monitoring studies, by nature, are correlative, and direct causative relationships are difficult to establish. The correlative nature of monitoring studies is especially important because natural populations are responding to a wide host of stressors (both natural and anthropogenic), and teasing apart cause–effect relationships is difficult, if not impossible. Similarly, the existence of true "reference sites" is debatable, especially for contaminants with widespread distributions like perchlorate (Chapter 1). Any investigation that compares reference sites to impacted sites is susceptible to possible confounding ecological factors. Monitoring studies also provide relatively little information about the mechanism of impact. If an impacted population appears to have lower abundances than a reference population, is it because of effects on individual growth, metabolism, reproductive capacity, or behavior? Finally, it is difficult to obtain funding to sustain a monitoring study for the length of time necessary to observe multiple generations of many organisms.

Mathematical models provide an alternative tool for addressing population-level effects of perchlorate. This chapter summarizes recent applications of population modeling within environmental toxicology with

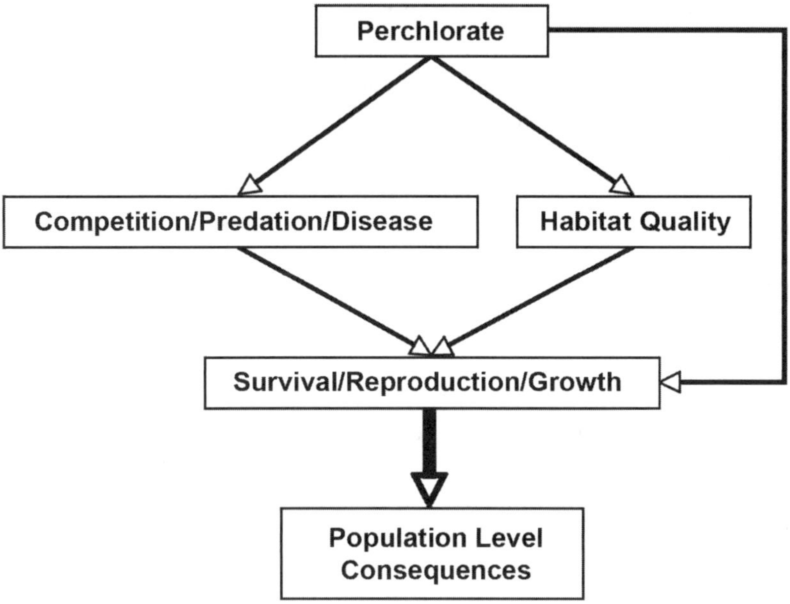

Figure 9-1 Possible effects of perchlorate that have population-level consequences. Direct effects include alterations of survival, growth, or reproduction. Indirect effects include modifications of the biotic and abiotic environment.

emphasis on approaches that may be especially relevant to studying perchlorate. Possible mechanisms of effect on populations are discussed, and a preliminary investigation of the possible population-level effects of perchlorate on amphibians is presented. Finally, recommendations for future research are made.

Mathematical Models

Mathematical modeling provides an efficient framework for evaluating the impact of contaminants on populations for a variety of reasons. First, models provide a cost-effective means of translating the effects of contaminants on individuals to effects on populations. Experiments designed to assess the impacts of contaminants on populations require time spans that cover multiple generations. The time, effort, and cost associated with executing such population-level experiments, especially on vertebrate species, are just too prohibitive. In addition, sensitivity analyses of input parameters provide insight into the specific variables that are important in determining impact, insights that may not be intuitively obvious based on knowledge about effects on individuals. For example, toxicity tests suggest that reproduction is the most sensitive endpoint of effect in the earthworm, *Lubricus rubellus* when exposed to copper (Klok and de Roos 1996). However, using a population model, Klok and de Roos (1996) found that the relatively small impact on growth was much more important in determining population responses than the relatively large impact of copper on reproduction. In this way, models can be used to test hypotheses about population behavior, given information about individual-level effects. In a similar fashion, "what if" scenarios can be run to examine possible impacts of a variety of mitigation strategies, future dosing regimes, or any other possible future scenarios. Finally, mathematical modeling provides a framework for synthesizing existing knowledge and for developing strategies for future research (McTavish et al. 1998).

Models can, however, introduce a level of complexity that reduces toxicologists' ability to provide concrete recommendations for policy decisions. Population modelers are faced with trying to balance model precision and realism with model simplicity, ease of interpretation, and ease of implementation. Population ecologists have tended to emphasize simplicity and tractability in order to capture the basic dynamics of populations and further theory development. Simplicity, however, frequently comes at the expense of realism. In order to test theories, models do not necessarily need to be validated against reality and, in fact, rarely are. The goal of risk assessment, however, is to provide a realistic estimate of the risk posed to natural populations by the presence of toxic substances. Simply put, population models that have been developed within ecology are not necessarily well suited to accomplishing this goal (Gross 1994; Hallam and Lassiter 1994).

Population-Level Modeling Approaches

Within the previous decade or so, significant strides have been made to synthesize population biology into an ecotoxicological framework. Although a thorough review of this work is beyond the scope of this chapter, several examples illustrate these approaches and highlight important considerations for addressing population-level effects of perchlorate. One of the most conceptually appealing ways to address population-level consequences is to design experiments to assess the impact of a contaminant directly on parameters of demographic models, especially the population growth rate, r (Walthall and Stark 1997; Forbes and Calow 1999, 2002). This approach has been proposed as a basis for ecological risk assessment; however, it requires costly, intensive, experimental designs that are not normally used in environmental toxicology and are impractical for many organisms.

Furthermore, in order to understand the impacts of any environmental contaminant on populations, it is important to consider the potentially confounding effects of population density (Forbes et al. 2001; Forbes and Calow 2002). "Density dependence" refers to the fact that as the number of individuals within a population increases, food and other resources become scarce. Hence, exposure to perchlorate at low densities may have very different effects than exposure at high densities. In fact, by removing intraspecific competitors, toxicants can have positive effects on the remaining individuals when toxicants occur at high densities (e.g., Moe et al. 2002). Forbes et al. (2001) found that it is difficult to predict the direction or magnitude of the interactive effects of density and toxicant exposure on population growth rates, even with knowledge about the life history dynamics of an organism. Therefore, it is critical that toxicologists take this density dependence into consideration when examining population-level endpoints. Nevertheless, most toxicological experiments are conducted in the absence of food limitations and ignore density dependence altogether.

Demographic models focus on the number of births and deaths within a population, while ignoring the specific mechanisms whereby contaminants possibly affect individual organisms. An alternative approach is to utilize models that explicitly consider the impact of contaminants on the physiology of individuals (Calow and Sibly 1990; Maltby 1999). Such approaches examine the direct (e.g., metabolic costs of detoxification) and indirect (lower energy assimilation) consequences of exposure to stress by calculating the amount of energy available for growth and reproduction, that is, the scope for growth. For example, Klok and de Roos (1996) incorporate individual-level energy budgets into a stage-structured population model of oligochaete worms to examine the impact of copper contamination. Such an approach would be valuable for understanding the impacts of perchlorate as information about its impact on energy budgets becomes available.

Another problem inherent in using many demographic models is that most do not consider individual-level variation (Hallam and Lassiter 1994). In contrast, most toxicological data focus on the variation among individuals in the endpoint of interest. In fact, individual-level variation is central to concepts such as the LD50. In order to address individual-level variation, one approach is to divide the population into a series of age or size classes, allowing the model to consider age-specific or size-specific effects. These "structured" populations are represented by matrix population models, and these models have a long history of use in the biological sciences (Caswell 2001). For example, Stark and Banken (1999) found that the age–stage structure of a population at the time of exposure influences the overall impact. Similarly, while examining a growth-regulating pesticide, Wennergren and Stark (2000) found that incorporation of age structure as well as pesticide persistence resulted in a delay in population growth as opposed to a simple reduction in the growth rate. A modification of this strategy that incorporates mechanistic aspects of effect is to divide the population into classes based upon physiological response. Although structured matrix models do allow for some variability among individuals, and are sometimes considered to be "individual-based," they assume that all individuals within a class are identical.

Individual-based models extend the division of a population and explicitly consider variation among individuals. In this sense, they become the extreme class-structured model (each individual is a class by itself). Such an approach is intuitively appealing because it allows for the incorporation of mechanistic models of contaminant effect within each of the individuals. Responses of every individual within the population are modeled and the impact of these responses on population-level endpoints assessed. However, the increasing complexity of such an approach results in a severe loss of analytical tractability and requires much more data for parameterization. Carlsen et al. (2004) suggest that the strength of individual-based models is the ability to conduct "what if" simulations to determine the relationship between contaminant release size and population dynamics.

In addition to considering individual-level variation, it is important to consider the effects of spatial heterogeneity. Contaminant concentrations, resource availability, and other environmental factors vary in space. In effect, this spatial heterogeneity can lead to individual-level variation in exposure and can be modeled via individual-based models. However, just as populations may be divided into classes based on age, size, or physiological status, populations may be divided into classes (i.e., subpopulations) based on their physical location (Spromberg et al. 1998; McLaughlin and Landis 2000). The only requirement is that the subpopulations be connected via migration. The spatial arrangement of these subpopulations can have significant impacts on the way in which the overall population (i.e., the metapopulation) responds to contaminants and other stressors. Furthermore, non-dosed subpopulations can experience ecologically significant effects because of the immigration of dosed individuals, or because of disruptions to migration caused by the contaminant (Chaumot et al. 2003a, 2003b).

Many contaminants vary continuously in space; however, modeling approaches that consider continuous variation in space (as opposed to simply dividing space into patches containing individual subpopulations) are lacking within ecotoxicology. Partial differential equations offer one method to address continuous variation, although these models are complex. In order to incorporate any degree of realism, they require a large number of parameters (which, in turn, requires large amounts of input data), and their use depends on numerical simulation techniques that necessitate high-performance computing. An alternative approach is to represent model dynamics via cellular automaton approximations. Cellular automata models represent space as a grid of habitat cells, and population dynamics within each of the cells can be modeled using demographic or individual-based methods. These approaches have been successful within ecology and epidemiology but have received little attention in ecotoxicology.

Stochasticity

For any of the approaches mentioned above, it is important to recognize that the biological processes that regulate population dynamics are not deterministic. That is, there is some element of chance (e.g., environmental conditions) that affects the birth and death processes within any population. This element of chance is called "environmental stochasticity." There is also some element of chance that is introduced as a result of having finite populations. For example, if we know that an individual within a population of 100 individuals has a 50% chance of producing 1 offspring, some years there will be 48, some 52, some 55, some 44, etc. That is, we do not necessarily expect there to be exactly 50 individuals (the deterministic prediction) produced every year. This variability is called "demographic stochasticity," and is particularly important in small populations. Inclusion of these sources of stochasticity within population models has been made possible only with the advent of relatively cheap computing power; however, inclusion of stochasticity can radically alter model behavior and resulting predictions (Renshaw 1991; Caswell 2001). When considering the effects of perchlorate on populations, it is important to recognize that there is some element of chance that determines

1) the concentration individuals may be exposed to,
2) the environmental conditions under which exposure may occur, and
3) the likelihood that exposure will lead to a measurable effect in an individual.

For example, fish whole-body concentrations of perchlorate observed in Texas were highly variable (Chapter 5).

Endocrine-Disrupting Chemicals

Perchlorate can affect thyroid structure and function and can result in altered endocrine function. Endocrine-disrupting chemicals (EDCs), especially organochlorines such as DDT, have received a considerable amount of attention. Nevertheless, very little is known about the effects of EDCs on populations (Sheffield et al. 1998), partly because the quantitative linkage between effects at the individual level and effects at the population level has received relatively little attention (Cooper and Kavlock 2001). In fact, the traditional toxicological paradigm may not apply to EDCs, even at the individual level (Crews et al. 2000). Instead, Crews et al. (2000) lay out a paradigm for addressing the potential effects of EDCs that recognizes the following:

1) exposure to EDCs is chronic,
2) effects may be latent,
3) a "lowest" dose with no adverse affects may not exist, and
4) EDCs rarely exist in isolation.

McTavish et al. (1998) lay out a modeling framework for examining the effects of EDCs on populations that incorporates many of these notions. Birth, death, and growth rates are represented as functions of age, contaminant dose, and population size. In this manner, aspects of individual variation and density dependence are captured. This framework also provides a mechanism for addressing the impacts of multiple contaminants, incorporating time delays in response, and for representing effects that can be passed on to offspring. This framework is especially relevant for examining possible population impacts of perchlorate.

Hypothesized Effects of Perchlorate on Populations

The effects of perchlorate on aspects of thyroid biology are well established. However, it is difficult to translate effects on thyroid hormone (TH) levels, thyroid function, or thyroid pathology to individual-level endpoints that are important at the population level. In particular, quantitative data on the effects of perchlorate on growth and reproduction or mortality are sparse. There is some indication of effects on population growth in the freshwater lamprey at 100 $mg \cdot kg^{-1}$ (ppm) and the freshwater hydra at 350 $mg \cdot kg^{-1}$; however, most evidence for population-level effects are based on conjecture that effects on the thyroid, pituitary, and brain could diminish survivorship and fecundity and diminish population production (draft EPA report).

Growth

Although THs are directly required for many aspects of tissue differentiation, they play an important permissive role in facilitating the effects of growth hormone and growth factors on somatic growth and body mass. Therefore, chronic exposure to perchlorate, sufficient to reduce serum TH concentrations, could negatively impact growth of newborn animals. There is some indication that perchlorate potentially inhibits growth within mammals (Chapter 4), and at environmentally relevant concentrations, perchlorate does slow down the rate of limb growth and metamorphosis in amphibians (Chapter 6; Goleman, Carr, Anderson 2002; Goleman, Urquidi, et al. 2002).

Reproduction

Perchlorate could possibly affect reproduction by altering the hypothalamic–pituitary–gonadotropic axis or one of its components, or by affecting the expression of steroid hormone receptors. There is some evidence that perchlorate reduces cocoon production within earthworms (Chapter 8). Moreover, there is substantial evidence that perchlorate can affect TH levels (Chapters 4 through 8), especially after prolonged exposures. However, the impact of those hormones on reproductive ability in vertebrates is unclear. THs are necessary for masculinization in anurans, and blocking THs can cause feminization (see Chapter 6). Alternatively, some studies in rats have shown effects of THs on reproduction (Dijkstra et al. 1996; Armada-Diaz et al. 2001), while others have not (York et al. 2001).

Metabolism

THs are required for the maintenance of body temperature in endotherms and in the regulation of metabolism. Perchlorate is not known to alter the distribution of energy among growth, reproduction, and maintenance by increasing maintenance costs. However, by affecting thyroid function, perchlorate may decrease total metabolism and result in lower energy available for somatic growth or reproduction (i.e., scope for growth). In birds, for example, such a reduction may affect the ability of precocial young to develop thermoregulatory ability (Chapter 5; McNabb and Olson 1996). It may also decrease an individual's ability to maintain metabolic homeostasis or cause deficiencies that would alter parental behavior.

THs have significant effects on thermogenesis (Silva 2003), and thermoregulation is a potentially important aspect of the metabolic effects of perchlorate that can be integrated into physiologically based models. TH is required for basal metabolic rate (BMR), and there are several mechanisms involved in the regulation of thermogenesis. For example, THs exert a general effect on ion homeostasis, which regulates cell metabolism. TH effects on heat production are critically important in the adjustment of newborn to cooler temperatures following birth, and disruption of the thermogenesis by perchlorate-induced TH depletion could lead to a reduced ability to maintain a normal body temperature. This, in turn, may lead to increased mortality in the newborn.

A second mechanism involves the permissive effect on the thermogenic action of catecholamines. Norepinephrine, which is secreted by sympathetic nerve endings, plays a predominant role in regulatory thermogenesis, especially nonshivering thermogenesis. Nonshivering thermogenesis is an important heat-generating mechanism in small mammals and hibernators. If hypothyroidism occurs during acute cold exposure, such as during ambient winter temperatures, it may result in hypothermia, which may be fatal.

Preliminary Analysis

As a first step in exploring possible population-level effects, we have examined the effects of perchlorate on amphibian species using an age-structured population model. There is increasing concern about the apparent decline in amphibian populations on a global scale (Houlahan et al. 2000), and chemical pollutants are included in the list of possible causes for such declines. Changes in UV-B irradiation as a result of depleted atmospheric ozone, pathogenic fungi, and climate change also have been implicated as possible causes for increasing egg mortality, leading to amphibian global declines. Population models afford an opportunity to examine these stressors, in isolation or in combination, at an ecologically relevant scale.

We used a model that was parameterized for the Western toad, *Bufo boreas* (Vonesh and De la Cruz 2002). The model assumes a generalized amphibian life cycle with synchronous breeding of adults in the spring. Juvenile (J) and adult (A) densities at time $t + 1$ are determined by multiplying densities at time t by a transition matrix. Hence,

$$\begin{bmatrix} J \\ A \end{bmatrix}_{t+1} = \begin{bmatrix} \sigma_j(1-P) & F(A) \\ \sigma_j(P) & \sigma_a \end{bmatrix} \begin{bmatrix} J \\ A \end{bmatrix}_t$$

where σ_j is the annual survival probability of juveniles, σ_a is the annual survival probability of adults, and P is the probability that a juvenile matures within a particular year. F(A) is a function that represents the production of new juveniles, per year, by adults, and it incorporates aspects of the sex ratio of adults, clutch size, egg survival, juvenile survival, and metamorph survival. Larval survival tends to be density dependent (Wilbur 1997); thus, within this production function, density-dependent larval survival is controlled by a density-dependent exponent.

Within amphibians, exposure to perchlorate can block masculinization (Goleman, Carr, Anderson 2002), thus leading to a change in the sex ratio of adults. It can also increase the time required to complete metamorphosis (Goleman, Carr, Anderson 2002; Goleman, Urquidi, et al. 2002), thus leading to a decrease in the percent of juveniles that undergo metamorphosis. We examined both of these potential impacts on amphibian populations using the limitation analysis approach of Vonesh and De la Cruz (2002). In essence, we looked at the percent reduction in adult equilibrium population size due to the effects of perchlorate as a population-level endpoint.

In the absence of data on perchlorate effects for the Western toad, we estimated perchlorate effect sizes from data on *Xenopus laevis* found in Goleman, Carr, and Anderson (2002). These data may or may not adequately estimate possible effects on the Western toad. We examined 4 different scenarios. First, perchlorate effects were modeled as a change in sex ratio (ρ), with control animals being 42% male, low-dose ($0.059\ mg \cdot kg^{-1}$ perchlorate) exposed animals being 33% male, and high-dose ($14\ mg \cdot kg^{-1}$ perchlorate) animals being 26% male. Second, we modeled the effect of perchlorate as a reduction in the number of juveniles undergoing metamorphosis (P). We used conservative estimates based on data for forelimb emergence and tail resorption, and we assumed a 25% reduction in metamorphosis in the low-dose group and a 50% reduction in the high-dose group. Third, we examined both of these effects simultaneously. Finally, we examined the effects of perchlorate in the presence of solar UV-B radiation, pathogenic fungi, and climate change (decreased water levels). We modeled the combined effects of these stressors as a 50% reduction in egg survival (σ_e), based on the data in Kiesecker et al. (2001).

Several important conclusions can be drawn from the model results, shown in Figure 9-2. First, the impact of a perchlorate-induced change in adult sex ratio depends heavily on the degree of density dependence within juveniles. At low levels of density dependence, altered sex ratios actually result in 10% to 17% higher equilibrium population sizes for the low- and high-dose groups, respectively. However, as density dependence increases, the increasing number of juveniles leads to increased competition and overall population decline. In contrast, possible reductions in the

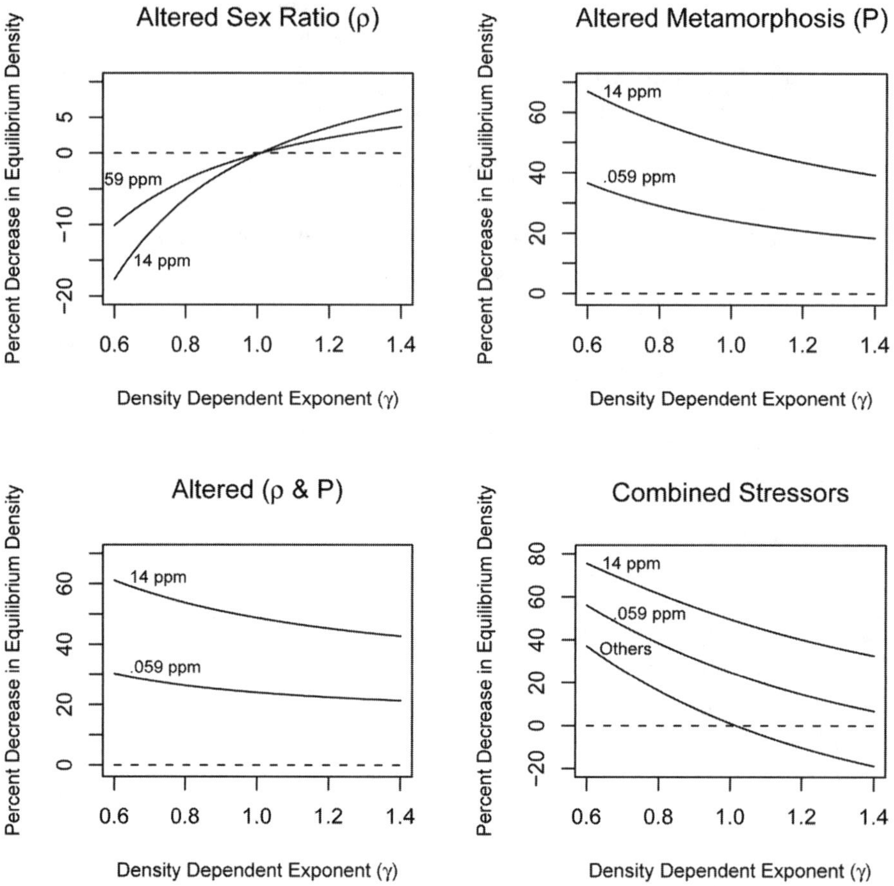

Figure 9-2 Percent reduction of equilibrium population size of adult frogs after exposure to perchlorate, based on an amphibian age-structured population model (Vonesh and De la Cruz 2002). Analyses were conducted over a range of likely density-dependent exponent values. Negative percent declines indicate increases in population sizes. Perchlorate effects were modeled as a change in sex ratio (top left), a change in the percent of juveniles who undergo metamorphosis (top right), and a change in both sex ratio and percent of juveniles who undergo metamorphosis (bottom left). Perchlorate effects are also investigated in the presence of other stressors (bottom right). The dashed line indicates control animals in the absence of all stressors.

number of juveniles undergoing metamorphosis as a result of perchlorate exposure leads to large reductions in population sizes at all levels of density dependence. The high-dose group goes from a 67% reduction at low levels of competition to a 39% reduction at high levels of competition. The low-dose group goes from a 36% reduction to an 18% reduction.

When both altered sex ratios and reductions in metamorphosis are combined, it is apparent that the strong negative effects of reductions in metamorphosis outweigh any positive effects of altered sex ratios. As a result, the overall effect of perchlorate is to reduce population sizes approximately 25% to 50% in the low- and high-dose groups, respectively. Moreover, the overall effect of perchlorate appears to be relatively independent of the degree of density dependence. This finding is in contrast to the synergistic effects of other stressors, which lead to increased egg mortality (Kiesecker et al. 2001). There, reductions in population sizes due to reduced egg survival at low levels of density dependence are ameliorated as density dependence increases. When perchlorate is added to the mix, however, populations experience reductions regardless of the level of density dependence. Nevertheless, the effect is much more severe (as high as 75% reduction in population size) at low levels of intraspecific competition.

Conclusions and Recommendations

Investigations into the effects of perchlorate on populations must not lose sight of the fact that it is ultimately the effect on individuals that leads to population-level consequences. Given the mechanisms hypothesized in this chapter, it is apparent that perchlorate may potentially alter individual growth, reproductive ability, or metabolism. Furthermore, the review of population-level modeling within environmental toxicology suggests that the conceptual and quantitative frameworks for using this individual-level information are available. For example, the preliminary model simulations indicate that by delaying metamorphosis, perchlorate can have profound effects on amphibian populations.

In order to make progress in understanding the effects of perchlorate on populations, toxicologists and population biologists need to collaboratively design laboratory experiments that will provide the data necessary to parameterize these mathematical models. Physiologically based models provide a mechanistically based approach that can be linked to a series of experiments that measure the metabolic effects of perchlorate exposure. Alternatively, the framework for EDCs appears to provide a more phenomenological approach that can be used to test a variety of hypotheses about perchlorate exposure, especially those that deal with transgenerational effects like the transfer of perchlorate to infants via breast milk (Chapter 4).

Ultimately, in order to understand the ecological effects of perchlorate, it is imperative to examine the impact of perchlorate at multiple levels of biological organization. Such a hierarchical approach facilitates a better understanding of the effects of stressors on ecosystems (Newman and Unger 2003), effects that may not be immediately obvious based on knowledge about toxicological effects on individuals. In fact, when we seek to understand the impact of perchlorate at a specific hierarchical level, information about the context and spatial–temporal constraints of effect is usually found at higher levels, whereas mechanistic information about effect is usually found at lower hierarchical levels (Hallam and Lassiter 1994; Pickett et al. 1994; Caswell 1996). Thus, individual-level effects provide the mechanism for understanding impacts on populations, while populations provide the context for understanding the effect of perchlorate on individuals.

References

Armada-Diaz L, Carvalho JJ, Breitenbach MM, Franci CR, Moura EG. 2001. Is the infertility in hypothyroidism mainly due to ovarian or pituitary functional changes? Braz J Med Biol Res 34:1209–1215.

Calow P, Sibly RM. 1990. A physiological basis of population processes: ecotoxicological implications. Funct Ecol 4:283–288.

Carlsen TM, Coty JD, Kercher JR. 2004. The spatial extent of contaminants and the landscape scale: an analysis of the wildlife, conservation biology, and population modeling literature. Environ Toxicol Chem 23:798–811.

Caswell H. 1996. Demography meets ecotoxicology: untangling the population level effects of toxic substances. In: Jagoe CH, editor. Ecotoxicology: a hierarchical treatment. Boca Raton (FL): Lewis.

Caswell H. 2001. Matrix population models: construction, analysis, and interpretation. Sunderland (MA): Sinauer Associates. p 722.

Chaumot A, Charles S, Flammarion P, Auger P. 2003a. Do migratory or demographic disruptions rule the population impact of pollution in spatial networks? Theoret Pop Biol 64:473–480.

Chaumot A, Charles S, Flammarion P, Auger P. 2003b. Ecotoxicology and spatial modeling in population dynamics: an illustration with brown trout. Environ Toxicol Chem 22:958–969.

Cooper RL, Kavlock RJ. 2001. Determining indicators of exposure and effects for endocrine disrupting chemicals (EDCs): an introduction. Human Ecol Risk Assess 7:971–978.

Crews D, Willingham E, Skipper JK. 2000. Endocrine disruptors: present issues, future directions. Quart Rev Biol 75:243–260.

Dijkstra G, de Rooij DG, de Jong FH, van den Hurk R. 1996. Effect of hypothyroidism on ovarian follicular development, granulosa cell proliferation and peripheral hormone levels in the prepubertal rat. Europ J Endocrinol 134:649–654.

Forbes VE, Calow P. 1999. Is the per capita rate of increase a good measure of population-level effects in ecotoxicology? Environ Toxicol Chem 18:1544–1556.

Forbes VE, Calow P. 2002. Population growth rate as a basis for ecological risk assessment of toxic chemicals. Phil Trans R Soc Lond 357:1299–1306.

Forbes VE, Sibly RM, Calow P. 2001. Toxicant impacts on density-limited populations: a critical review of theory, practice, and results. *Ecol Appl* 11:1249–1257.

Goleman WL, Carr JA, Anderson TA, 2002. Environmentally relevant concentrations of ammonium perchlorate inhibit thyroid function and alter sex ratios in developing *Xenopus laevis*. Environ Toxicol Chem 21:590–597.

Goleman WL, Urquidi LJ, Anderson TA, Kendall RJ, Smith EE, Carr JA. 2002. Environmentally relevant concentrations of ammonium perchlorate inhibit development and metamorphosis in *Xenopus laevis*. Environ Toxicol Chem 21:424–430.

Gross LJ. 1994. Limitations of reductionist approaches in ecological modeling: model evaluation, model complexity, and environmental policy. In: Kendall RJ, Lacher TE, editors. Wildlife toxicology and population modeling: integrated studies of agroecosystems. Boca Raton (FL): Lewis. p 576.

Hallam TG, Lassiter RR. 1994. Individual-based mathematical modeling approaches in ecotoxicology: a promising direction for aquatic population and community ecological risk assessment. In: Kendall RJ, Lacher TE, editors. Wildlife toxicology and population modeling: integrated studies of agroecosystems. Boca Raton (FL): Lewis. p 531–542.

Houlahan JE, Findlay CS, Schmidt BR, Meyer AH, Kuzmin SL. 2000. Quantitative evidence for global amphibian population declines. Nature 404:752–755.

Kendall RJ, Lacher TE. 1994. Wildlife toxicology and population modeling: integrated studies of agroecosystems. SETAC Foundation for Environmental Education, Proceedings of the Ninth Pellston Workshop; 1990 Jul 22–27; Kiawah Island (SC). Boca Raton (FL): Lewis.

Kiesecker JM, Blaustein AR, Belden LK. 2001. Complex causes of amphibian population declines. Nature 410:681–683.

Klok C, de Roos AM. 1996. Population level consequences of toxicological influences on individual growth and reproduction in *Lumbricus rubellus* (Lumbricidae, Oligochaeta). Ecotoxicol Environ Saf 33: 118–127.

Maltby L. 1999. Studying stress: the importance of organism-level responses. Ecol Appl 9:431–440.

McLaughlin JF, Landis WG. 2000. Effects of environmental contaminants in spatially structured environments. In: Albers PH, Heinz GH, Ohlendorf HM, editors. Environmental contaminants and terrestrial vertebrates: effects on populations, communities, and ecosystems. Pensacola (FL): Society of Environmental Toxicology and Chemistry (SETAC). p 351.

McNabb FMA, Olson JM. 1996. Development of thermoregulation and its hormonal control in precocial and altricial birds. Poult Avian Biol Rev 7:111–125.

McTavish K, Stech H, Stay F. 1998. A modeling framework for exploring the population-level effects of endocrine disruptors. Environ Toxicol Chem 17:58–67.

Moe SJ, Stenseth NC, Smith RH. 2002. Density-dependent compensation in blowfly populations give indirectly positive effects of a toxicant. Ecology 83:1597–1603.

Newman MC, Unger MA. 2003. Fundamentals of ecotoxicology. Boca Raton (FL): Lewis. p 458.

Pastorok RA, Ginzburg LR. 2002. Conclusions and recommendations. In: Pastorok RA, Bartell SM, Ferson S, Ginzburg LR, editors. Ecological modeling in risk assessment. Boca Raton (FL): Lewis. p 302.

Perchlorate Environmental Contamination: Toxicological Review and Risk Characterization (2002 External Review Draft). U.S. Environmental Protection Agency, Office of Research and Development, National Center for Environmental Assessment, Washington Office, Washington, DC, NCEA-1-0503, 2002.

Pickett ST, Kolasa J, Jones CG. 1994. Ecological understanding. San Diego (CA): Academic Pr.

Renshaw E. 1991. Modelling biological populations in space and time. Cambridge (NY): Cambridge Univ Pr.

Sheffield SR, Matter JM, Rattner BA, Guiney PD. 1998. Fish and wildlife species as sentinels of environmental endocrine disruptors. In: Kendall RJ, Dickerson RL, Giesy JP, Suk WA, editors. Principles and processes for evaluating endocrine disruption in wildlife. Pensacola (FL): Society of Environmental Toxicology and Chemistry (SETAC). p 515.

Silva JE. 2003. The thermogenic effect of thyroid hormone and its clinical implications. Ann Intern Med 139:205–213.

Spromberg JA, Bettina MJ, Landis WG. 1998. Metapopulation dynamics: indirect effects and multiple distinct outcomes in ecological risk assessment. Environ Toxicol Chem 17:1640–1649.

Stark JD, Banken JAO. 1999. Importance of population structure at the time of toxicant exposure. Ecotoxicol Environ Saf 42:282–287.

Suter II GW, Barnthouse LW. 2000. Modeling toxic effects on populations: Experience from aquatic studies. In: Albers PH, Heinz GH, Ohlendorf HM, editors. Environmental contaminants and terrestrial vertebrates: effects on populations, communties, and ecosystems. Pensacola (FL): Society of Environmental Toxicology and Chemistry (SETAC). p 351.

Vonesh JR, De la Cruz O. 2002. Complex life cycles and density dependence: assessing the contribution of egg mortality to amphibian declines. Oecologia 133:325–333.

Walthall WK, Stark JD. 1997. Comparison of two population-level ecotoxicological endpoints: the intrinsic (rm) and instantaneous (ri) rates of increase. Environ Toxicol Chem 16:1068–1073.

Wennergren U, Stark JD. 2000. Modeling long-term effects of pesticides on populations: beyond just counting dead animals. Ecol Appl 10:295–302.
Wilbur HM. 1997. Experimental ecology of food webs: complex systems in temporary ponds. Ecology 78:2279–2302.
York RG, Brown WR, Girard MF, Dollarhide JS. 2001. Two-generation reproduction study of ammonium perchlorate in drinking water in rats evaluates thyroid toxicity. Int J Toxicol 20:183–197.

ten

Ecological Risk Assessment

Philip N Smith, Jody Wireman, James A Carr, Ernest E Smith, Ronald J Kendall

Introduction

The purpose of this chapter is to assimilate available perchlorate-related information presented within this book in the context of an ecological risk assessment (ERA). Although specific screening values are not developed, perchlorate-related ecotoxicology information is reviewed to highlight data, concepts, and uncertainties regarding this contaminant, and likely ecological receptors that should be considered as screening values are derived. Here we attempt to integrate environmental perchlorate concentrations in various abiotic and biotic ecosystem components with qualitative and quantitative effects information derived from recent laboratory studies to identify potential risks to ecosystems. Field studies examining perchlorate exposure and related effects are also considered, though they are often complicated because of the occurrence of confounding factors and/or extraneous stressors.

As described in Chapters 1 and 2, perchlorate is an oxidizing anion found as a contaminant in both ground and surface waters as a result of the dissolution of perchlorate salts (ammonium, potassium, magnesium, and sodium). Perchlorate salts are highly water soluble and dissociate in water to release perchlorate. The perchlorate anion is largely unreactive in dilute aqueous solutions at temperatures and pHs found in nature (Urbansky 1998). Additionally, perchlorate is unlikely to adsorb to soils because it is an anion and its absorption to minerals is weak and reversible (Urbansky et al. 2001). Once in water, perchlorate can persist for many years and can easily migrate with its water source.

Perchlorate Ecotoxicology. Ronald J Kendall and Philip N Smith, editors.

ISBN-10: 1-880611-90-2 ISBN-13: 978-1-880611-90-6

Perchlorate competitively inhibits iodide uptake into thyroid gland follicular cells by binding to the sodium (Na^+)-iodide (I^-) symporter (NIS), thereby reducing synthesis of the thyroid hormones (THs) tetraiodothyronine (T_4) and triiodothyronine (T_3) (see Chapter 3 for a thorough discussion of the mechanism of action). The premise of perchlorate ecotoxicology studies described throughout this book is based on the identified effects of perchlorate on thyroid function observed in humans and laboratory rats, and on the possibility that these effects could occur in aquatic and wildlife species, because THs play important roles in the development of most species (Kendall et al. 1998).

Considerable scientific and regulatory debate and effort have gone into the evaluation of human health risks associated with perchlorate; however, limited information has been included in evaluations of potential ecological risks. This chapter considers these and other recent perchlorate-related toxicity studies unavailable for inclusion in the United States Environmental Protection Agency (USEPA) screening-level ERA for perchlorate completed in 1998 and updated in 2002. Data that became available after the "fall [of] 2001" were not incorporated into USEPA risk evaluations. Since then, a number of studies have been published that fill data gaps identified in the USEPA reports. More data now exist to characterize perchlorate risks, including information on aquatic toxicity and bioaccumulation, plant uptake, terrestrial mammal and invertebrate exposures and effects, and trophic transfer. For example, when the USEPA developed its perchlorate screening values, limited information was available regarding aquatic and terrestrial vertebrates, invertebrates, and vegetation. Previously, aquatic risks were believed to occur at approximately 4 to 10 $mg \cdot L^{-1}$ (Parsons Inc. 2001). However, recent data suggest that ammonium perchlorate can inhibit metamorphosis in frogs (Goleman, Urquidi, et al. 2002) and adversely affect thyroid histopathology in larval frogs (Tietge et al. 2005) and fishes (Mukhi et al. 2005) at concentrations lower than 0.02 $mg \cdot L^{-1}$. Potassium perchlorate concentrations in soils as low as 8.6, 18, and 107 $mg \cdot kg^{-1}$ can reduce biomass of lettuce, tomato, and oats, respectively, by 50% (Adema and Henzen 1989). Until recently, terrestrial toxicity information has been limited to laboratory rodent studies conducted for the human health risk assessment and one acute invertebrate study that evaluated perchlorate toxicity in earthworms (*Eisinea fetida*).

In addition, few field studies evaluating potential effects from real-world exposures on contaminated sites were available to the USEPA in the fall of 2001. This chapter includes field study results that may eventually be evaluated in the context of laboratory results and, possibly, to determine potential effects of perchlorate on population dynamics. However, each preceding chapter identifies remaining data gaps that contribute uncertainty to attempts at defining the ecological risk posed by perchlorate.

Problem formulation

We offer a modified version of a conceptual site model (CSM) developed by the USEPA in their draft risk assessment released in 2002 to illustrate which pathways and taxa are potentially impacted by the environmental occurrence of perchlorate. This modified CSM is used as the starting point for this evaluation of the potential ecological risks of perchlorate (Figure 10-1). Modifications include the addition of likely ecological receptors and alternative routes of exposure. Based on our knowledge about the occurrence of perchlorate in the environment (i.e., completed path-

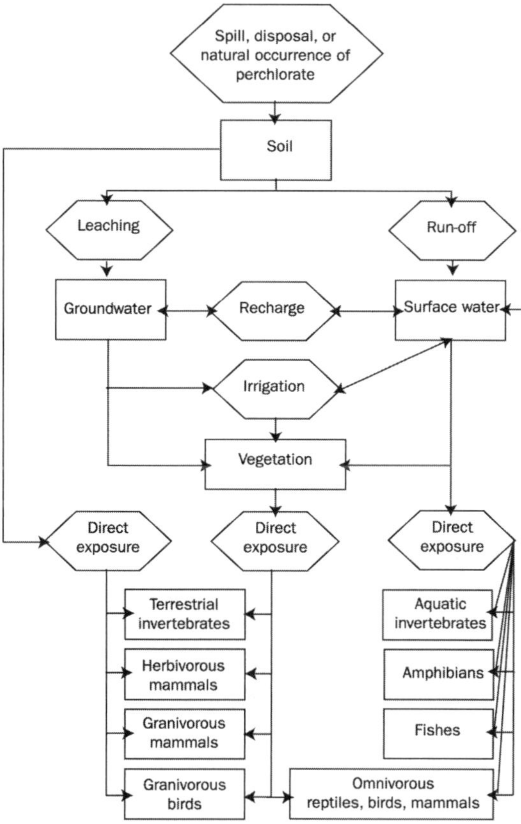

Figure 10-1 Conceptual model for perchlorate exposures to ecological receptors. Although not depicted in this conceptual model, additional exposure to some organisms may occur transplacentally.

ways) and on associated toxicity information, freshwater species appear to have the greatest potential for exposure and, subsequently, for effects. The pathway analyses approach suggests that terrestrial species should be included in an ERA for perchlorate to include species that utilize both land and water resources, especially as a major food source (e.g., amphibians, reptiles, and rodents). Previously, ecological exposures were considered to occur primarily through receptor contact with contaminated water sources. However, recent information on uptake and accumulation of perchlorate in plants suggests that animal exposure via contaminated plant-stuffs may have been underestimated. Research on ecological exposure to date has mirrored our knowledge about perchlorate occurrence, and here we attempt to synthesize and present this information.

Potential ecological exposure pathways associated with perchlorate are depicted in the CSM (Figure 10-1). Sources for perchlorate contamination are summarized in Chapter 1. To date, it appears that the historical disposal of unused perchlorate salts and spills or releases as part of manufacturing processes have led to the highest concentrations of perchlorate in the environment. Additionally, contributions from the use of perchlorate in commercial, military, and agricultural settings, as well as other potential sources, are known to occur but have not been fully evaluated. Perchlorate has been detected in a few saltwater areas, but saltwater systems have not yet been examined thoroughly. It is difficult to discern whether levels in saltwater are elevated relative to freshwater sources and, if so, whether adverse effects would potentially result from such exposures. In addition, the potential for adverse effects resulting from perchlorate exposures is unclear because of the coexistence of high levels of iodide in saltwater.

Conceptual model

This CSM provides the direct relationship between those species that have been evaluated and where they appear in relation to completed pathways (a more thorough description for fish is included in Chapter 5). Since the USEPA developed its CSM, we have gained information from field and laboratory studies that has allowed us to expand the CSM to include organisms at higher trophic levels. As noted above and in Chapter 2, plants are exposed to perchlorate through water transport processes. Wildlife are assumed to have negligible exposures via the inhalation pathway and through dermal contact (with the notable exception of aquatic organisms). The major route of exposure to perchlorate for wildlife is through ingestion of contaminated water, food items (particularly vegetation grow in the presence of perchlorate), and soil. Additionally, respiratory and dermal exposure should be considered for aquatic species. Perchlorate exposure can be understood semiquantitatively through field study evaluations that are inclusive of all pathways (Figure 10-1).

Perchlorate meets the criteria for designation as an endocrine-disrupting chemical (USEPA 1997), and as such, can potentially alter endocrine function in all exposed ecological receptors, with consequential effects primarily observed in developing organisms. However, long-term low-level exposure may potentially impact adult organisms that require normal TH synthesis for development and reproduction, ultimately affecting individual fitness and population health. The following 4 assessment endpoints are addressed in this chapter:

1) Survival, health, and reproduction of freshwater organisms potentially exposed to perchlorate in water, sediment, and food items (i.e., plants, invertebrates, fish, and amphibians).
2) Survival, health, and reproduction of terrestrial invertebrates potentially exposed to perchlorate in soils, sediments, pore water, and vegetation.
3) Survival, health, and reproduction of birds using waterways and potentially exposed to constituents in freshwater, sediment, soil, and seeds.
4) Survival, health, and reproduction of mammalian species using freshwater areas and potentially exposed to perchlorate in freshwater, soil, plants, and prey items.

Analysis

Ecological exposure

The ability to measure the presence of perchlorate and its fate and transport through media have improved with the advancement of analytical and cleanup methods. As mentioned in previous chapters, perchlorate is persistent in surface and ground waters and has the ability to enter other abiotic media, and biotic media at various trophic levels. Environmental occurrence, partitioning, and observed concentration ranges for perchlorate in water, soil, and sediment are described in detail within Chapters 1 and 2.

Nearly every freshwater and terrestrial habitat is a potential area where perchlorate exposures could occur. It is also possible for all species whose home and feeding ranges include all or portions of these areas to become exposed to perchlorate. Two prominent studies that first described the occurrence and distribution of perchlorate in ecosystems were conducted by Parsons Inc. (2001) and Smith et al. (2001). It is important to note that these studies were conducted over a short duration and co-location was not accomplished in all sampling (e.g., some water samples may have been collected a day or 2 prior to or after a fish sample was collected).

Parsons Inc. (2001) conducted an occurrence survey of perchlorate residuals in various biotic tissues and abiotic media at six different perchlorate-contaminated sites across the United States. Among the 965 samples collected were surface water, groundwater, soil, sediment, sediment pore water, aquatic and terrestrial birds, fish, aquatic and terrestrial reptiles, amphibians, terrestrial mammals, aquatic and terrestrial insects, other aquatic and terrestrial invertebrates, and aquatic and terrestrial plants. The objective was to cover a wide range of potential sources, pathways, and receptors. Average measured perchlorate concentrations ranged from 0.7 to 58 $mg \cdot kg^{-1}$ for all media (these concentrations were not corrected for nondetected values, which was a limitation of the extraction and analytical methods used). Of all the media sampled, soil and terrestrial vegetation typically exhibited the highest perchlorate concentrations and frequency of detection, followed by aquatic vegetation, surface water, sediment pore water, and sediments. Perchlorate was only occasionally present at detectable concentrations in organisms at higher trophic levels.

Smith et al. (2001) also reported perchlorate measurements in aquatic organisms in perchlorate-contaminated streams and ponds near the Longhorn Army Ammunition Plant. Perchlorate concentrations ranged from 0.8 to 2.0 $mg \cdot kg^{-1}$ in aquatic insects, below detection (BD) to 0.2 $mg \cdot kg^{-1}$ in fish, BD to 0.6 $mg \cdot kg^{-1}$ in frogs, BD to 2.3 $mg \cdot kg^{-1}$ in mammals, and 0.6 to 5557 $mg \cdot kg^{-1}$ in vegetation. As in the Parsons Inc. study (2001), many of the highest perchlorate concentrations were measured in terrestrial plants, followed by water, sediment, and aquatic plants. Concentrations of perchlorate in aquatic fauna were lower than or similar to those in water at most sites, indicating a general lack of bioconcentration.

The Parsons Inc. (2001) and Smith et al. (2001) studies documented the movement of perchlorate through abiotic media and exposure among a wide range of ecological receptors. Of particular interest was the observation of perchlorate residues in organisms that were not intimately associated with contaminated surface water sources indicating alternative exposure pathways. The highest concentrations of perchlorate were observed in vegetation in both studies, which provided evidence that vegetation was an important source of exposure for terrestrial organisms.

In other species, whole body tissue concentrations were less than the concentration present in the co-located matrix evaluated (i.e., food, soil, or water). Therefore, the bioconcentration factors (BCFs) and bioaccumulation factors (BAF) have been less than 1. The exception has been the BCF for clams, which was approximately 1.85 (Dean et al. 2004). It should also be noted that perchlorate tends to accumulate in the thyroid gland and other organs. Analysis of these organs might be of greater utility in determining perchlorate exposures.

Bioconcentration and bioaccumulation information is generally assimilated for a risk assessment to assess whether trophic transfer can or is occurring. Perchlorate has been observed to accumulate in various plant species through ion uptake channels in both field and laboratory studies (Nzengung et al. 1999; Susarla et al. 2000; Urbansky et al. 2000; Ellington et al. 2001; Smith et al. 2001; Aken and Schnoor 2002; Tan et al. 2004; Yu et al. 2004). BCFs have ranged from 40 to 628 on a wet-weight basis for the transpiration tissues of food crops, such as soybean leaves and pods, tomato leaves, alfalfa, wheat stems and heads. Perchlorate was also detected in edible portions (fruits and seeds) of several garden plants, though a lower degree of bioconcentration was typically observed (BCF ranged from 0.5 to 20). In general, results from the food crop studies were consistent with the BCF (dry weight basis) calculated for aquatic vegetation of 164 and 266 (Tan, Jackson, et al. 2004).

There is a wide array of available habitat types where perchlorate is known or likely to be present which, in turn, host a wide variety of aquatic species and wildlife. Thus, complete ecological exposure pathways are described in general terms, but are based on the following common scenarios:

- exposure of freshwater aquatic resources (plants, invertebrates, and fish) to perchlorate in freshwater
- exposure of freshwater aquatic resources (invertebrates and fish) to perchlorate in sediment
- exposure of freshwater aquatic and terrestrial resources (invertebrates, fish, amphibians, reptiles, birds, and mammals) to perchlorate in sediment, freshwater, and the food chain at source areas
- exposure of terrestrial resources (plants, invertebrates, birds, mammals, reptiles, and amphibians) to perchlorate in soils

Analysis of ecological effect

The ecological effects assessment identifies the toxicological properties associated with perchlorate. It determines the type of effect that could result to ecosystem components if perchlorate exposures are excessive, and identifies which benchmarks provide a measure of the potential for ecological effects.

Mammals

Exposure to varying concentrations of perchlorate and its associated effects have been documented in several mammalian species (Siglin et al. 2000; Thuett et al. 2002a, 2002b; Baldridge et al. 2004; Cheng et al. 2004; York et al. 2004; Isanhart et al. 2005). Perchlorate administered to Sprague Dawley rats in a subchronic toxicity test for 14 or 90 d in drinking water at dosages ranging from 0.01 to 10 $mg \cdot kg^{-1} \cdot d^{-1}$ ammonium perchlorate significantly increased thyroid weight and thyroid histopathology in the form of follicular cell hypertrophy with microfollicle and colloid depletion (Siglin et al. 2000). Male Wistar rats dosed with ammonium perchlorate for 14 d at 10 $mg \cdot kg^{-1} \cdot d^{-1}$ were reported to exhibit decreased TH plasma concentrations (Siglin et al. 2000). Chronic exposure (730 d) of Wistar rats to perchlorate resulted in increased thyroid weight and some follicular hyperplasia (Kessler and Kruskemper 1966) with 36% of the treated animals developing benign thyroid tumors after 2 years. Exposure of rats to ammonium perchlorate salt (0.0 to 30 $mg \cdot kg^{-1} \cdot d^{-1}$) from 14 d before cohabitation to gestation day 21 resulted in increased maternal thyroid weight (both absolute and relative) in the 30 $mg \cdot kg^{-1} \cdot d^{-1}$ treatment group (York et al. 2003). Evaluation of the offspring's thyroid weight and histopathology in Sprague Dawley rats exposed to perchlorate during gestation and lactation (York et al. 2004) revealed pronounced colloid depletion and increased hypertrophy at 0.1 $mg \cdot kg^{-1} \cdot d^{-1}$ and hyperthrophy and hyperplasia at 3.0 $mg \cdot kg^{-1} \cdot d^{-1}$ perchlorate salt.

Deer mice (*Peromyscus maniculatus*) offspring from breeding pairs dosed continuously with perchlorate (0 to 117 $mg \cdot L^{-1}$ ammonium perchlorate salt) had significantly decreased colloidal area and decreased numbers of follicles per unit area that contained colloid and increased total plasma T_4. Body and organ weight data for perchlorate-treated deer mice for the 117 $\mu g \cdot L^{-1}$ dose group were altered in a manner opposite to that of the 117 $ng \cdot L^{-1}$ and 117 $mg \cdot L^{-1}$ treatment groups. It appears as though in deer mice perchlorate has a biphasic effect, that is, concentrations around 117 $\mu g \cdot L^{-1}$ have a different effect than do the lower and higher concentrations of 117 $ng \cdot L^{-1}$ and 117 $mg \cdot L^{-1}$. The body weights of 117 $mg \cdot L^{-1}$ treated females were higher than that of controls and of other treatments, whereas the male body weights were similar irrespective of treatment. Liver weights of deer mice exposed to perchlorate (Thuett et al. 2002a) at 117 $mg \cdot L^{-1}$ were significantly higher than those in the 117 $\mu g \cdot L^{-1}$ treatment group for litter 2 pups. Testicular morphological changes have been observed in ammonium perchlorate treated deer mice. An increase in seminiferous tubule diameter of the testes of deer mice was proposed to be a secondary effect resulting from the decrease in THs (Thuett et al. 2002b).

In Prairie voles (*Microtus ochrogaster*), an herbivorous rodent species, plasma thyroxine concentrations decreased in a dose-dependent manner following exposure for 51 d to 0 to 10 $mg \cdot kg^{-1} \cdot d^{-1}$ ammonium perchlorate in water (Isanhart et al. 2005). However, there were no significant differences in plasma T_4 concentrations in the perchlorate treatment groups voles when compared to control animals. However, mean thyroidal T_4 concentrations of the 10.0 $mg \cdot kg^{-1}$ treatment group was significantly lower than that of the control group. Mean thyroid gland T_4 concentration was 464 ± 182.05 μg T_4/thyroid gland and mean plasma T_4 concentration was 2.69 ± 0.41μg/dL in voles exposed to 0.75 mg perchlorate$\cdot kg^{-1} \cdot d^{-1}$ for 180 d (Isanhart et al. 2005).

In summary, thyroid gland morphology and TH concentrations were consistently altered in all mammalian species evaluated to date. In addition, developmental toxicity has been identified in the lab rat and deer mice following gestational and lactational exposure. Based on the range of observed perchlorate concentrations in water and vegetation at contaminated sites (Parsons Inc. 2001; Smith et al. 2001), there is a possibility that mammalian wildlife could be exposed to high enough concentrations of perchlorate that the exposure could result in ecologically significant biological effects (Figure 10-2). However, there are no field data currently available to support this assertion. To the contrary, Smith et al. (2004) were unable to identify perchlorate-related changes in THs among raccoons (*Procyon lotor*) inhabiting a contaminated site in east Texas, which may indicate that higher trophic level species or species that eat a variety of food sources may not be exposed to sufficient perchlorate concentrations to elicit an effect. Alternatively, ambient iodide concentrations in available food resources and/or the mammalian capacity for long-term storage of THs may ameliorate episodic or chronic exposure to perchlorate.

Avian

Gentles et al. (2005) evaluated the transfer of perchlorate into eggs from Northern bobwhite (*Colinus virginianus*) hens exposed to ammonium perchlorate in water. Egg perchlorate concentrations increased with increasing exposure. At 1.18 $mg \cdot L^{-1}$ ammonium perchlorate in the drinking water, hens deposited 0.6 $\mu g \cdot g^{-1}$ of perchlorate ion in their eggs, at 11.75 $mg \cdot L^{-1}$, egg perchlorate was 3.8 $\mu g \cdot g^{-1}$ and at 117.5 $mg \cdot L^{-1}$, eggs contained 7.2 $\mu g \cdot g^{-1}$ perchlorate ion. There were no observed changes in egg dimensions or in numbers of eggs laid in any of the ammonium perchlorate treatment groups compared to control hens drinking distilled water (Gentles et al., in press). Northern bobwhite quail hens drinking the highest perchlorate concentration (117.5 $mg \cdot L^{-1}$ ammonium perchlorate) used by Gentles et al. (in press) showed significant decreases in colloid area and significant increases in cell height in the thyroid gland.

Assessment of several thyroid variables indicated that thyroid function is altered (based on decreased thyroidal hormone content; primarily thyroidal T_4) at much lower perchlorate exposures and at shorter exposure times than those that result in organismal hypothyroidism as indicated by circulating TH concentrations. Thus, thyroidal T_4 content was significantly decreased in 21-d bobwhite quail embryos from eggs with a concentration of ≥0.5 $\mu g \cdot g^{-1}$ ammonium perchlorate and in hatchlings from eggs with ≥50 $mg \cdot L^{-1}$ ammonium perchlorate. However, egg concentrations of up to 300 $\mu g \cdot g^{-1}$ did not result in organismal hypothyroidism in these embryos or hatchlings. Mallard (*Anus platyrhincos*) embryos were much less sensitive to perchlorate than quail embryos; thyroidal T_4 content was decreased at ≥20 $\mu g \cdot g^{-1}$ in 24-d mallard embryos (McNabb et al. 2002; McNabb, Larsen, Pooler 2004)

Bobwhite quail chicks show alterations in thyroid function within two weeks of ammonium perchlorate exposure; thyroidal T4 is decreased at ≥0.05 $mg \cdot L^{-1}$ ammonium perchlorate but organismal hypothyroidism is only present at ≥500 $mg \cdot L^{-1}$. Surprisingly, if perchlorate exposure is continued, chicks show adaptation of their thyroid function at some concentrations. Three different concentration-dependent patterns occur. From 0.05 to 5 $mg \cdot L^{-1}$ ammonium perchlorate, there is full recovery or restoration (i.e., adaptation) of thyroidal-T_4 stores to control levels. Whereas, at 50 and 250 $mg \cdot L^{-1}$, there is partial restoration of thyroidal-T_4. At the highest levels tested (500 to 4000 $mg \cdot L^{-1}$ ammonium perchlorate) thyroidal-T_4 stores remain severely depleted (McNabb, Jang, Larsen 2004). Organismal hypothyroidism (as indicated by plasma T_4) only occurs consistently in chicks drinking ≥500 $mg \cdot L^{-1}$ ammonium perchlorate for 2 weeks or ≥250 $mg \cdot L^{-1}$ for 8 weeks. Mallard ducklings have only been studied in relation to two week perchlorate exposures. At this time point their thyroid function is much less sensitive to perchlorate than is the case for quail; duckling thyroidal T4 is decreased significantly at ≥5.0 $mg \cdot L^{-1}$ ammonium perchlorate (McNabb et al. 2003; McNabb 2004). In adult bobwhites, the effects of perchlorate are slower to develop, increase with time, and higher perchlorate exposures are tolerated than in chicks. In addition there is no indication of thyroid adaptation to perchlorate, at least up to 8 weeks, the longest exposure studied.

Amphibians

Several papers have reported on the effects of environmentally relevant concentrations of perchlorate on metamorphosis in *Xenopus laevis*. Metamorphosis was delayed at 0.005 to 0.1 $mg \cdot L^{-1}$ perchlorate and inhibited at 0.147 $mg \cdot L^{-1}$ (70-d exposure period) (Goleman, Urquidi, et al. 2002). In a similar study, McDaniel (2004) observed that 0.050 mg perchlorate/L inhibited metamorphosis. Metamorphosis also was inhibited at 0.250 $mg \cdot L^{-1}$ in *X. laevis* (Tietge et al. 2005), although a much shorter exposure period was used compared to the longer exposure durations examined in Goleman et al. (Goleman, Carr, Anderson 2002; Goleman, Urquidi, et al. 2002) and McDaniel (2004). Goleman, Carr, and Anderson (2002) reported that the effects of ammonium perchlorate were reversible when tadpoles were exposed to ammonium perchlorate for 70-d period and then allowed a 28-d nontreatment recovery period. Thyroid follicle cell hypertrophy was observed at 0.056 $mg \cdot L^{-1}$ (Goleman, Carr, Anderson 2002) and at 0.016 $mg \cdot L^{-1}$ (Tietge et al. 2005) in *X. laevis* exposed to either ammonium or sodium perchlorate, respectively. Colloid depletion in response to perchlorate also has been reported for *X. laevis* larvae exposed to 0.016 $mg \cdot L^{-1}$ sodium perchlorate for 14 d (Tietge et al. 2005). Perchlorate concentrations between 0.056 and 14 $mg \cdot L^{-1}$ alter sex ratios in developing *X. laevis* in a concentration-dependent manner, resulting in more females (Goleman, Carr, Anderson 2002).

Field studies indicate follicle cell hypertrophy and colloid depletion in chorus frog (*Pseudacris triseriata*) tadpoles inhabiting a pond containing approximately 10 mg perchlorate/L relative to stage-matched tadpoles from a reference site (Carr et al. 2003). Thyroid follicle cell height was significantly larger in adult male cricket frogs (*Acris crepitans*) inhabiting perchlorate contaminated streams (USACE 2004). Cricket frogs inhabiting ponds contaminated with an average of 0.010 to 0.032 $mg \cdot L^{-1}$ had significantly more follicle cell hypertrophy than adult male frogs collected from streams containing 0.005 mg perchlorate/L or less. The degree of follicle cell hypertrophy was positively and significantly correlated with perchlorate concentration (USACE 2004; Theodorakis et al. 2005). In these studies perchlorate values represented an average value from several measurements taken over an 18 month period in 2001 and 2002.

Laboratory studies also evaluated perchlorate exposures and its affect on oxidative stress and susceptibility to UV radiation (McDaniel 2004). It was generally found that perchlorate exposure increased susceptibility to UV-induced DNA damage and mortality at concentrations as low as 50 $\mu g \cdot L^{-1}$.

Fishes

Potential for exposures is most readily seen via colloidal T_4 ring intensity evaluations, which indicate the presence of perchlorate at 0.011 mg·L^{-1} (Mukhi et al. 2005). In addition, thyroidal effects were observed at about 0.1 mg·L^{-1} in terms of hypertrophy (Bradford et al., in press; Mukhi et al. 2005). In addition, thyroidal effects (thyroid follicle hyperplasia, hypertrophy, colloid depletion, and increased angiogenesis) were observed in mosquitofish (*Gambusia holbrooki*) exposed to 0.1 mg·L^{-1} sodium perchlorate for 30 d (Bradford et al., in press; Mukhi et al. 2005). Thyroidal effects (hyperplasia, hypertrophy, colloid depletion) in the field were seen in fish from sites contaminated with perchlorate concentrations less than 80 µg·L^{-1} (Theodorakis et al. 2005). These data indicate that, at environmentally relevant concentrations (<1 mg·L^{-1}), perchlorate has the potential to induce perturbations in thyroidal homeostasis, as measured by the above endpoints. Fitness parameters have not been demonstrated for fish, although the importance of TH in piscine physiology suggests potential adverse physiological effects on pathways influenced by THs. Such endpoints include growth and embryonic development, osmoregulation, immune function, and gonadal development (see Chapter 7 for review).

There is equivocal evidence for fitness effects of perchlorate in fish (Brown et al. 2004). Growth and pubertal development of sub-adult zebrafish (*Danio rerio*) were not affected by exposure to ammonium perchlorate at 0 to 11.48 mg·L^{-1} (Mukhi et al. 2005). However, increased incidences of macrophage aggregates occurred in zebrafish at 18 mg·L^{-1} and mosquitofish at 92 mg·L^{-1} (Capps et al. 2004). The latter may not have environmental relevance based on concentrations typically found near perchlorate sources. There were no adverse effects on fecundity and growth in mosquitofish exposed to perchlorate for 6 weeks. Other adverse effects cannot be ruled out at environmental concentrations.

Invertebrates

Studies conducted previously suggested that perchlorate (as sodium perchlorate) is not particularly lethal to earthworms, with acute (14 d) and subchronic (28 d) LC50s of 4450 mg·kg^{-1} and 5071 mg·kg^{-1}, respectively. Anderson et al. report similar observations in Chapter 8 of this volume. The acute (14 d dermal contact test) LC50s for sodium perchlorate and ammonium perchlorate to *E. fetida* were 4150 mg·kg^{-1} and 4140 mg·kg^{-1}, respectively. More realistic exposure scenarios (14 d sand tests) and slightly longer exposure times (21 d) produced data suggesting that perchlorate was slightly more toxic under these conditions; the acute (14 d sand test) LC50 for sodium perchlorate to *E. fetida* was

2550 $mg \cdot kg^{-1}$ and the subchronic (21 d) LC50 for sodium perchlorate to *E. fetida* was 500 $mg \cdot kg^{-1}$.

Reproductive toxicity tests reported by Anderson et al. were more variable but indicated that cocoon production was more sensitive to perchlorate toxicity. The acute (14-d sand tests) EC50 for sodium perchlorate to *E. fetida* reproduction (cocoon production) was 260 $\mu g \cdot g^{-1}$. In the treatments that reproduced, cocoon production was slightly higher at 1 $\mu g \cdot g^{-1}$ than at 1 $ng \cdot g^{-1}$. Highest cocoon production was observed in the Controls. The subchronic (21-d sand tests) EC50 for sodium perchlorate to *E. fetida* reproduction (cocoon production) was 1.3 $\mu g \cdot g^{-1}$. The subchronic (28 d artificial soil tests) EC50 for sodium perchlorate to *E. fetida* reproduction (cocoon production) was 350 $\mu g \cdot g^{-1}$. In the treatments that reproduced, cocoon production was slightly higher in the Control and 10 $\mu g \cdot g^{-1}$ treatments than in the 100 $ng \cdot g^{-1}$ and 100 $\mu g \cdot g^{-1}$ treatments.

Risk Characterization and Uncertainty

An ecological risk characterization is done to evaluate the evidence linking perchlorate with potential adverse ecological effects. This link is established by combining the ecological exposure assessment and ecological effects assessment through quantitative or qualitative evaluations, or both. Potential risk to ecological receptors is determined on the basis of both quantitative and qualitative evaluations.

Laboratory information has been supplemented with field data when available as evidence that an effect is related to a perchlorate exposure (i.e., lines of evidence). The lines of evidence can be drawn from indications of effects from multiple species and through comparison of field and laboratory results. For example, lines of evidence may be drawn from the identification of perchlorate residuals in various media that exceed screening levels or perchlorate concentrations in whole-body tissue that exceed effects-based tissue residue benchmarks.

Conclusions

It is possible that mammalian receptors may be exposed to concentrations of perchlorate at contaminated sites such as the Las Vegas Wash, Nevada (Smith et al. 2004), such that adverse effects might occur. Although no direct perchlorate-related effects have been observed in wild mammals, field data are not sufficient to dismiss the possibility that adverse effects might occur on heavily contaminated sites. Likewise, under worst case conditions, birds can obtain perchlorate doses that lead to endocrine disruption and potential target organ effects. Immature mammals, including those in utero, may be exposed gestationally, or through lactation and may be at greatest risk for adverse effects from alterations in TH-related processes. However, TH storage capacity and sufficient iodide intake may offset potential effects perchlorate may have on adult mammals. Young birds appear more sensitive than adults, but chicks exhibit some adaptation to perchlorate with continued low to moderate level exposures, which may also be protective of wild birds. Further, there is considerable interspecies variability in response to perchlorate in birds which increases uncertainty in assessing ecological risk.

Amphibians have proven to be the most sensitive indicators of environmental perchlorate contamination to date. Environmentally relevant concentrations of perchlorate have been observed to alter metamorphosis in amphibians. Altered amphibian metamorphosis would likely result in ecological effects, especially in those species inhabiting ephemeral ponds and having rapid developmental rates. However, there are no empirical data linking perchlorate exposure to population-level effects in amphibians. Amphibian responses to perchlorate depend upon the timing of exposure, both in the length of exposure and the period during development when exposure occurs. Adult frogs are not as sensitive to perchlorate exposure as are developing frogs; and tadpoles at later stages of development are not as sensitive to the antimetamorphic effects of perchlorate, as are tadpoles at earlier stages of development.

Native fish populations are not expected to experience acute toxic effects of perchlorate, but chronic exposure to perchlorate in the laboratory and the field has resulted in thyroid gland structural and physiological perturbations as well as alterations in TH concentrations. Perchlorate does not bioconcentrate in fish, and to date, evidence for detrimental effects of perchlorate on ecologically relevant fitness parameters at environmentally relevant concentrations remains equivocal. Similarly, there is no evidence of perchlorate-induced effects such as growth, survival, and reproduction in fish at environmentally relevant concentrations.

Perchlorate is not lethal to earthworms at environmentally relevant soil concentrations, but chronic perchlorate exposure can reduce earthworm mass and thus, perhaps fitness. Further, environmentally relevant concentrations of perchlorate may affect earthworm reproduction.

Complete characterization of ecological risks requires a discussion of the uncertainties inherent in the assumptions used to estimate exposures and risks. This section focuses on uncertainties pertaining to the information presented throughout this book with the intention of highlighting areas in need of additional research.

Uncertainty and data gaps

The data for perchlorate occurrence in the United States has largely focused on protection of drinking water resources, with relatively few investigations considering ecological exposures. Some important studies have been conducted at specific locations to expand basic information on perchlorate detection and quantitative measurements of ecosystem fate and exposure. There may be more useful information on perchlorate ecological exposure and fate in site investigation reports that are not readily available to the scientific community. Clearly, there is a pressing need for ecological exposure information at specific sites and for expanding the information base available for general use in perchlorate ecotoxicology.

Use of more robust analytical methods to examine distribution at levels less than 4 $\mu g \cdot L^{-1}$ is already beginning to further our knowledge of ecological exposure of perchlorate, including the potential for a background concentration in some regions in the nanogram per liter range. Future research on ecotoxicology would benefit greatly by a review of existing data and standardized analytical methods with lower detection limits.

Basic research also is needed on the natural formation and accumulation of perchlorate (Bao and Gu 2004). Some effort has begun to develop stable isotope fingerprinting methods to attempt to differentiate between manmade and naturally formed perchlorate. There are legitimate questions on the conditions under which perchlorate formed in nature may accumulate in the environment, and whether this source of perchlorate may constitute a significant ecological exposure.

There has also been discussion on whether ion chromatography methods are best suited to screen for the presence of perchlorate in matrices other than water and whether these more complex samples should be reanalyzed or confirmed using mass spectroscopy (LC/MS or IC/MS). If this is the case, it may not be possible to link some of the results (especially complex matrices) from either laboratory or field studies.

Iodide and perchlorate competition and interaction

Since iodide and perchlorate compete for the same transport protein, the NIS, the amount of iodide that is in the blood stream and available to the thyroid should in theory alter the degree that perchlorate binds to the NIS and alters thyroid function. In isolated cell preparations, the mammalian NIS protein has a higher affinity for perchlorate than for iodide, which means that both anions do not compete equally for the transport protein and that given equimolar concentrations of both substrates, the NIS will transport more perchlorate than iodide. Given that the NIS is also located in several extrathyroid tissues, changes in iodide kinetics would also be expected to influence perchlorate transport across these tissues.

Terrestrial vertebrates obtain iodide through their diet, but there are virtually no studies examining how changes in dietary iodide availability affect the kinetics of perchlorate and iodide transport by thyroid or other tissues. Based on data collected in neonatal rats, Clewell et al. (2003) modeled the impact of dietary iodide intake on iodide levels in serum, thyroid, and other tissues. Clewell et al. (2003) hypothesized that only tissues which express the NIS would be expected to show changes in iodide content with change in iodide availability in the diet. Their data indicate that increasing dietary iodide in drinking water to concentrations as great as 8 $mg \cdot L^{-1}$ does not appreciably affect thyroid iodide concentrations. Although this model needs to be confirmed with empirical data, the preliminary data suggest that changes in dietary intake of iodide do not appreciably alter thyroidal NIS activity, although the effects on perchlorate transport specifically have not been tested.

Aquatic vertebrates acquire iodide through their food and may also obtain iodide through the skin or respiratory surfaces such as gills. Work in two different laboratories indicates that environmental iodide availability can dramatically affect the ability of perchlorate to disrupt thyroid-sensitive aspects of development. Sparling et al. (2003) showed that addition of small amounts of iodide to the rearing medium negated the inhibitory effects of perchlorate on metamorphosis in tree frog tadpoles (*Hyla versicolor*). Similarly, Hu et al. (2003) demonstrated that equimolar concentrations of iodide in rearing medium completely blocks perchlorate's effects on metamorphosis in the African clawed frog, *X. laevis*.

Vegetation

Plant BCFs are influenced by season and growth stage of the plants. Vegetative uptake of perchlorate in terrestrial species appears to be a function of both exposure concentration and length of exposure (Tan, Anderson, Jones, et al. 2004). Uptake of perchlorate by terrestrial plant species cannot be viewed as permanent sequestration. Leaf litter data (Tan, Anderson, Jones, et al. 2004) suggest that perchlorate could be released once plant leaves senesce in the fall and may be viewed as a mechanism of perchlorate exposure to lower trophic systems and a means of enlarging the exposure area (this also has the potential to influence aquatic and terrestrial invertebrates and vertebrates). Use of BCFs is not appropriate in systems in which the source of perchlorate is variable. Little data are available on the kinetics of perchlorate uptake but certainly equilibrium time will be on the order of weeks (Tan, Anderson, Jackson 2004; Tan, Jackson, et al. 2004). Additionally, plants with deep or extensive root systems may draw water from multiple sources complicating the calculation of steady state tissue concentrations. In general, BCF application must be used with caution unless detailed site data are available on the spatial and temporal variability of perchlorate in relation to distribution of edible plants.

Critical needs for research related to the uptake of perchlorate in plants also remain. These include the mechanism of uptake into the plant root and the associated molecular control systems, the rate of uptake and influencing factors, and more distribution data especially for fruits. Also of interest is the continuing analytical development of methods for detection of perchlorate at low levels in difficult matrices (e.g., organs, plants). Finally, given the importance of the element iodine in relation to the thyroidal effects of perchlorate, better methods may be needed for analysis of the environmentally relevant forms of iodine as well and a better understanding of perchlorate uptake in plants in relation to iodine (all forms) uptake.

Toxicological data

Uncertainties in toxicological data can also influence the reliability of risk management decisions. As with many contaminants, data on perchlorate toxicity in aquatic and wildlife receptors are limited. In addition, the usefulness of existing toxicity information in assessing ecological impacts is constrained by several factors. Most wildlife toxicity information is generated by laboratory studies with selected test species. These studies frequently evaluate laboratory cultured animals under controlled conditions, with few tests involving native wildlife. Basic toxicity information can be extrapolated to native species in the wild, but consideration must be given to the species involved and specific site conditions. Attempts were made throughout this book to locate studies using native species and, when necessary and appropriate, supplement this information with results from studies that utilized laboratory species.

Extrapolations from tested species to other species, whose sensitivities to perchlorate may not be homogeneous, results in uncertainty. Additionally, effects of perchlorate-induced thyroid alterations on physiological (e.g., immunity, osmoregulation, temperature acclimation, bioenergetics), behavioral (e.g., predator–prey interactions, competition, swimming performance, sexual behaviors), population, or community level effects have not been evaluated. An example is the uncertainty associated with potential effects on juvenile development and behavior which have not been fully evaluated.

Mammals. Dose–response evaluations at environmentally relevant concentrations are needed to establish the association with thyroid disruption effects. An understanding of maternal exposure and milk transfer to neonatal and developing offspring are needed for more realistic risk characterization and assessment. Definitive hypothesis driven evaluations of the potential metabolic and thermoregulatory effects of perchlorate in various species are warranted as are studies examining the effects of perchlorate on cellular and molecular mechanism of action, for example the association of perchlorate with transport membrane proteins such as pendrin.

Birds. The pattern of development of thyroid function is dramatically different in precocial compared with altricial bird species (Chapter 5, Figure 5-1). Only precocial species have been evaluated, thus far, because they are species traditionally evaluated by the USEPA. However, the effects from the different amounts of circulating THs between precocials and altricials, suggest that altricials should be further evaluated.

Additional studies are needed to define the measurement protocols that should be used to establish how thyroid disruption or its downstream effects should be evaluated in birds. Specifically, altered thyroid histology should be compared with measurements of thyroidal TH content because these two thyroid variables appear to be the most sensitive to thyroid disruption. Studies of avian food items and water sources indicate that the levels of exposure required for affecting thyroid histology and depleting TH stores occur in some situations where environmental perchlorate contamination is present. The occurrence and consequences of a hypothyroid state on bird development, health, survival and reproduction need to be studied in the context of environmentally relevant exposures. These studies need to be done on both precocial and altricial birds because of the differences in the patterns of thyroid, organ system and thermoregulatory ontogeny in these two developmental modes. Additional studies at low, environmentally relevant perchlorate concentrations and for longer time periods are needed to fully assess the nature of exposures that cause organismal hypothyroidism and downstream target organ effects. Studies of perchlorate, in the context of the background literature on hypothyroid effects on developing avian systems (specifically, nervous, skeletal, and muscular tissues as well as feather molting) need further investigation.

Neurodevelopmental effects influencing behavior, survival and reproduction in nestlings and in adults need to be investigated in birds. Current research on the effects of THs on central nervous system development in mammals, as well as recent studies of TH effects on the plasticity of neuronal cell turnover in the song nucleus of the avian brain, suggest the need for investigations in this area. Because THs are important in the control of metabolism, the effects of perchlorate-induced hypothyroidism on the development of thermoregulation should be investigated in both altricial and precocial birds. These studies may be especially fruitful in very small birds with very high metabolic rates, and may especially affect thermoregulatory development in nestling altricial young both directly and indirectly through the high energetic demands on parental care of nestlings. The interactions between environmental iodine availability and perchlorate effects in birds need to be investigated. Iodide availability can alter responses to perchlorate in humans and laboratory mammals; high iodide availability is known to lessen the effects of perchlorate low iodide availability enhances perchlorate effects (see Chapter 3) but there is no available information about this interaction in birds.

Amphibians. There remain many uncertainties concerning the mechanism of perchlorate action in amphibians. Regarding the effects of UV and perchlorate exposure, the applicability of data from artificial UV sources to situations with natural sunlight should be addressed. The NIS gene has not been fully cloned or sequenced in any amphibian. Thus, controlled studies on the kinetics of the isolated amphibian NIS are lacking. Determining the kinetic parameters of this transporter in amphibians will allow researchers to determine how exposure endpoints such as blood levels of perchlorate might interfere with iodide transport. Moreover, the physiological role(s) of extrathyroidal NIS proteins need to be better understood in order to characterize the most likely routes of perchlorate exposure in amphibians. The role that environmental iodide plays in determining the response of amphibians to perchlorate needs to be better elucidated. As described earlier in this chapter, the relative availability of iodide to the organism should be considered when examining perchlorate effects. Although some research indicates that perchlorate may interfere with normal gonadal differentiation in frogs (Goleman, Carr, Anderson 2002), the implications of perchlorate exposure during gonadal differentiation for reproductive success and population stability are unknown. Although population modeling holds promise for testing the population level effect of perchlorate, multigenerational exposure studies should also be conducted to gather empirical data on the biological significance of perchlorate-induced changes in sex ratio. As mentioned in Chapter 6, a true LOEC for perchlorate effects on thyroid function in amphibians has not been reported. In every study in which it has been examined, and at every concentration that has been tested, perchlorate exposure causes follicle cell hypertrophy in larval *X. laevis*. Future studies should include perchlorate concentrations sufficiently low to bracket a LOEC for perchlorate on thyroid histology in tadpoles. Finally, most studies examining the sublethal effects of perchlorate exposure to date have been carried out in *X. laevis*, a species that is not native to North America. The degree to which data on perchlorate exposure can be extrapolated from *X. laevis* to other amphibian species is unclear.

Fish. More information is needed about the effects of perchlorate on fitness parameters in fishes. Although no adverse effects of environmentally relevant concentrations of perchlorate on growth or reproduction have been demonstrated in any fish species, the available data are insufficient to make generalizations, especially for the early life stages that are typically more sensitive to disruption by environmental contaminants. Accordingly, early life-stage and full life-cycle or partial life-cycle tests are necessary for adequate analysis of risk.

In mammals, TH disruption has been implicated in male reproductive effects, including effects on sperm production, testicular function, and steroid synthesis (Buzzard et al. 2000; Krassas and Perros 2003; Maran 2003), but such effects have not been examined in fish.

Although it is known that perchlorate may affect the amount of THs produced by adult fish, the potential for perchlorate to cause transgenerational effects in fish has yet to be investigated. Embryonic fish require THs for development but generally do not synthesize the hormones themselves until after hatching. THs in embryos are supplied via maternal transfer to the eggs. If TH synthesis is disrupted in the mother, it may affect her ability to transfer THs to her eggs and thus embryonic development may be affected. Although perchlorate may affect the amount of THs produced by adult fish especially after prolonged exposures, the effects of perchlorate-induced thyroid endocrine disruption on maternal transfer of THs to eggs, and potential consequential effects on fish embryo development, are matters that have yet to be resolved.

Another area for future research is examination of alternative physiological fitness endpoints of perchlorate exposure. For example, embryonic neuronal development is controlled by THs (Denver et al. 1997); thus, it is possible that perchlorate-induced thyroid disruption could result in neurobehavioral effects, but this has not been investigated. In addition, perchlorate exposure and thyroid suppression have been found to suppress the immune system in amphibians (Rollins-Smith 2001), and THs have been found to affect immune function in fish as well (Harris and Bird 2000), but the potential effect of perchlorate on fish immune function has not been explored. Finally, possible effects of environmentally relevant concentrations of perchlorate on other thyroid-responsive physiological endpoints (e.g., lipid metabolism, temperature acclimation, osmoregulation, pigmentation, mitochondrial function, embryo or larval development) and expression of thyroid-responsive and fitness-related genes merits further attention. In particular, studies on thyroid-responsive gene expression could be used to provide 1) a method for distinguishing between pathological effects and compensatory responses and 2) tools for studying thyroid disruption in embryos, larvae, and early juveniles, which are too small for conventional TH assays.

In a related light, interactions between THs and other endocrine, exocrine, and paracrine signaling pathways are an important consideration. For example, since growth hormone gene expression may be controlled or influenced by TH status, does perchlorate affect growth hormone levels? Other potential effects could be less direct via cross talk between TH and sex hormone, corticosteroid, and retinoid signaling pathways.

An additional data gap is how perchlorate exposure is affected by other naturally occurring and anthropogenic chemicals. The most relevant naturally occurring chemical is iodide or other oxidative states of iodine (e.g., iodate), because availability of iodine in the water could be critical for determining the magnitude of perchlorate-induced effects. In terms of anthropogenic chemicals, it is not known if perchlorate affects the toxicity, environmental chemistry and fate, toxicokinetics, or detoxification of other toxicants, and vice versa. Potential mechanisms for such interactive effects could include redox reactions between chemicals, thyroid-mediated changes in metabolism of chemicals, or nonthyroidal effects on fish physiology. Finally, the toxicity and behavior of perchlorate in complex mixtures is unknown.

In terms of PBPK and TH models, more work needs to be done on the testing, validation, refinement, and sensitivity analysis. In particular, inclusion of components for tissue- or compartment-specific peripheral deiodination, metabolism, excretion, and storage (especially storage in follicle colloids) would greatly enhance utility, realism, and accuracy of thyroid models. The models could also be expanded to include molecular, receptor- or ligand-based, toxicodynamic, and computational approaches.

Finally, more field studies are needed to properly assess the ecotoxicological risk of perchlorate. These may include microcosm or mesocosm studies or field surveys to further characterize environmental fate of perchlorate and pathways of exposure for organisms in aquatic systems. Such studies would provide answers to questions that are critical for assessing ecological risk of perchlorate. For example, to what extent do each of the potential pathways and routes of exposure play in the exposure and effects of perchlorate in fish? Also, what are the possible indirect effects of perchlorate on fish populations? Examples of indirect effects include

1) perturbations of aquatic food webs due to toxic effects on plankton or macroinvertebrates;
2) differential sensitivities to perchlorate among species; and
3) interactions between perchlorate and natural variables or stressors (e.g., pathogens, salinity, pH, temperature, dissolved oxygen, suspended sediment, ultraviolet radiation, food availability). Other manipulative or survey-type studies are needed to determine the extent of perchlorate exposure on reproduction and development in field situations, and the effects of perchlorate contamination on populations or communities of fish. Until these issues are resolved, it will be difficult to complete and ERA of the effects of perchlorate in natural environments.

Invertebrates. The number of studies evaluating perchlorate effects on aquatic and terrestrial invertebrates has increased during the last couple of years. However, the database on invertebrates is still relatively small. The data accumulated to date suggest that perchlorate is not lethal to earthworms at environmentally relevant soil concentrations (exceptions being spills or point sources). Perchlorate does appear to affect earthworm reproduction at concentrations that would have environmental relevance. In addition, chronic perchlorate exposure may reduce earthworm weight, which could influence survival over the long term. It is recommended that endpoints other than lethality be examined in order to determine the subtle effects that perchlorate may have on terrestrial invertebrates.

Population-level effects. Investigations into the effects of perchlorate on populations must not lose sight of the fact that it is ultimately the effect on individuals that leads to population level consequences. Given the hypothesized mechanisms above, it is apparent that perchlorate may potentially alter individual growth, reproductive ability, or metabolism. Furthermore, the review of population level modeling within environmental toxicology suggests that the conceptual and quantitative frameworks for utilizing this individual level information are available. Indeed, in order to make progress in understanding the effects of perchlorate on populations, toxicologists and population biologists need to collaboratively design experiments that will provide the data necessary to parameterize these mathematical models. Physiologically based models provide a mechanistically based approach that can be linked to a series of experiments which measure the metabolic effects of perchlorate exposure. The framework for EDCs appears to provide a more phenomenological approach that can be used to test a variety of hypotheses about perchlorate exposure, especially those that deal with transgenerational effects like the transfer of perchlorate to infants via breast milk (Chapter 4).

Ultimately, in order to understand the ecological effects of perchlorate, it is imperative to examine the impact of perchlorate at multiple levels of biological organization. Such a hierarchical approach facilitates a better understanding of the effects of stressors on ecosystems (Newman and Unger 2003), effects that may not be immediately obvious based on knowledge about toxicological effects on individuals. In fact, when seeking to understand the impact of perchlorate at a specific hierarchical level, information about the context and spatial or temporal constraints of effect are usually found at higher levels whereas mechanistic information about effect is usually found at lower hierarchical levels (Hallam and Lassiter 1994; Pickett et al. 1994; Caswell 1996). Thus, individual-level effects provide the mechanism for understanding impacts on populations while populations provide the context for understanding the effect of perchlorate on individuals.

References

Adema DM, Henzen L. 1989. A comparison of plant toxicities of some industrial chemicals in soil culture and soilless culture. Ecotoxicol Environ Saf 18:219–229.

Aken BV, Schnoor JL. 2002. Evidence of perchlorate (ClO4-) reduction in plant tissues (poplar tree) using radio-labeled 36ClO4-. Environ Sci Technol 36:2783–2788.

Baldridge MG, Stahl RL, Gerstenberger SL, Tripoli V, Hutz RJ. 2004. In utero and lactational exposure of Long-Evans rats to ammonium perchlorate (AP) disrupts ovarian follicle maturation. Reprod Toxicol 19:155–161.

Bao H, Gu B. 2004. Natural perchlorate has a unique oxygen isotope signature. Environ Sci Technol 38:5073–5077.

Bradford CM, Carr JA, Rinchard J, Theodorakis C. 2005. Perchlorate affects thyroid function in eastern mosquitofish (*Gambusia holbrooki*) at environmentally relevant concentrations. Environ Sci Technol 39(14):5190–5195.

Brown SB, Adams BA, Cyr DG, Eales JG. 2004. Contaminant effects on the teleost fish thyroid. Environ Toxicol Chem 23:1680–1701.

Buzzard JJ, Morrison JR, O'Bryan MK, Song Q, Wreford NG. 2000. Developmental expression of thyroid hormone receptors in the rat testis. Biol Reprod 62:664–669.

Capps T, Mukhi S, Rinchard J, Theodorakis CW, Blazer VS, Patiño R. 2004. Exposure to perchlorate induces the formation of macrophage aggregates in the trunk kidney of zebrafish and mosquitofish. J Aquat Anim Health 6:145–151.

Carr JA, Urquidi LJ, Goleman WL, Hu F, Smith PN, Theodorakis CW. 2003. Ammonium perchlorate disruption of thyroid function in natural amphibian populations: assessment and potential impact, In: Linder G, editor. Multiple stressor effects in relation to declining amphibian populations. West Conshohocken (PA): American Soc for Testing and Materials. ASTM STP 1443. p 131–142.

Caswell H. 1996. Demography meets ecotoxicology: untangling the population level effects of toxic substances. In: Jagoe CH, editor. Ecotoxicology: a hierarchical treatment. Boca Raton (FL): Lewis.

Cheng Q, Perlmutter L, Smith PN, McMurry ST, Jackson WA, Anderson TA. 2004. A study on perchlorate exposure and absorption in beef cattle. J Agric Food Chem 52:3456–61.

Clewell RA, Merrill EA, Yu KO, Mahle DA, Sterner TR, Fisher JW, Gearhart JM. 2003. Predicting neonatal perchlorate dose and inhibition of iodide uptake in the rat during lactation using physiologically-based pharmacokinetic modeling. Toxicol Sci 74:416–436.

Dean KE, Palacheck RM, Noel JM, Warbritton R, Aufderheide J, Wireman J. 2004. Development of freshwater water-quality criteria for perchlorate. Environ Toxicol Chem 23:1441–1451.

Denver RJ, Pavgi S, Shi YB. 1997. Thyroid hormone-dependent gene expression program for *Xenopus* neural development. J Biol Chem 13: 8177–8188.

Ellington JJ, Wolfe NL, Garrison AW, Evans JJ, Avants JK, Teng Q. 2001. Determination of perchlorate in tobacco plants and tobacco products. Environ Sci Technol 35:3213–3218.

Gentles A, Surles J, Smith EE. Evaluation of adult quail and egg production following exposure to perchlorate-treated water. Environ Toxicol Chem 24(8):1930–1934.

Goleman WL, Carr JA, Anderson TA, 2002. Environmentally relevant concentrations of ammonium perchlorate inhibit thyroid function and alter sex ratios in developing *Xenopus laevis*. Environ Toxicol Chem 21:590–597.

Goleman WL, Urquidi LJ, Anderson TA, Kendall RJ, Smith EE, Carr JA. 2002a. Environmentally relevant concentrations of ammonium perchlorate inhibit development and metamorphosis in *Xenopus laevis*. Environ Toxicol Chem 21:424–430.

Hallam TG, Lassiter RR. 1994. Individual-based mathematical modeling approaches in ecotoxicology: a promising direction for aquatic population and community ecological risk assessment. In: Kendall RJ, Lacher TE, editors. Wildlife toxicology and population modeling: integrated studies of agroecosystems. Boca Raton (FL): Lewis. p 531–542.

Harris J, Bird DJ. 2000. Modulation of the fish immune system by hormones. Vet Immunol Immunopathol 77:163–176.

Hu F, Gentles A, Goleman WL, Carr JA. 2003. Developmental stage and environmental iodide influence the antimetamorphic effects of perchlorate. Society of Environmental Toxicology and Chemistry 24th Annual Meeting; 2003 Nov 9–13; Austin, TX.

Isanhart JP, McNabb FM, Smith PN. 2005. Effects of perchlorate exposure on resting metabolism, peak metabolism, and thyroid function in the prairie vole (*Microtus ochrogaster*). Environ Toxicol Chem 24:678–684.

Kendall RJ, Dickerson RL, Giesy JP, Suk WP. 1998. Principles and processes for evaluating endocrine disruption in wildlife. Pensacola (FL): Society of Environmental Toxicology and Chemistry (SETAC).

Kessler FJ, Kruskemper HL. 1966. Experimental thyroid tumors caused by many years of potassium perchlorate administration. Klin Wochenschr 44:1154–1156.

Krassas GE, Perros P. 2003. Thyroid disease and male reproductive function. J Endocrin Invest 26:372–380.

Maran RR. 2003. Thyroid hormones: their role in testicular steroidogenesis. Arch Androl 49:375–388.

McDaniel LN. 2004. Ultraviolet radiation-induced genotoxicity in *Xenopus laevis* exposed to sodium perchlorate [MS thesis]. Lubbock (TX): Texas Tech Univ.

McNabb FMA, Jang DA, Larsen CT. 2004. Does thyroid function in developing birds adapt to sustained ammonium perchlorate exposure? Toxicol Sci 82:106–113.

McNabb FMA, Larsen CT, Pooler PS. 2004. Ammonium perchlorate effects on thyroid function and growth in bobwhite quail chicks. Environ Toxicol Chem 23:997–1003.

McNabb FMA, Maher RD, Jones JE, Jang DA. 2003. Ammonium perchlorate effects on developing and adult thyroid function in bobwhite quail and mallard ducks. Soc Integr Comp Biol Abstr. 43.1, p 247.

Mukhi S, Carr JA, Anderson TA, Patino R. 2005. Development and validation of new biomarkers of perchlorate exposure in fishes. Environ Toxicol Chem 24:1107–1115.

Newman MC, Unger MA. 2003. Fundamentals of ecotoxicology. Boca Raton (FL): Lewis. p 458.

Nzengung VA, Wang CH, Harvey G. 1999. Plant-mediated transformation of perchlorate into chloride. Environ Sci Technol 33:1478–1479.

Parsons Inc. 2001. Scientific and technical report for perchlorate biotransport investigation: a study of perchlorate occurrence in selected ecosystems. Interim final report. Brooks Air Force Base (TX): Air Force Institute for Environment, Safety and Occupational Health Risk Analysis.

Pickett ST, Kolasa J, Jones CG. 1994. Ecological understanding. San Diego (CA): Academic Pr.

Rollins-Smith LA. 2001. Neuroendocrine-immune system interactions in amphibians. Immunol Res 23:273–280.

Siglin JC, Mattie DR, Dodd DE, Hildebrandt PK, Baker WH. 2000. A 90-day drinking water toxicity study in rats of the environmental contaminant ammonium perchlorate. Toxicol Sci 57:61–74.

Smith PN, Theodorakis CW, Anderson TA, Kendall RJ. 2001. Preliminary assessment of perchlorate in ecological receptors at the Longhorn Army Ammunition Plant (LHAAP), Karnack, Texas. Ecotoxicology 10:305–313.

Smith PN, Yu L, McMurry ST, Anderson TA. 2004. Perchlorate in water, soil, vegetation, and rodents collected from the Las Vegas Wash, Nevada, USA. Environ Pollut 132:121–127.

Sparling DW, Harvey G, Nzengung V. 2003. Interaction between perchlorate and iodine in the metamorphosis of *Hyla versicolor*. In: Linder G, editor. Multiple stressor effects in relation to declining amphibian populations, ASTM STP 1443. ASTM International, West Conshohocken (PA): American Soc for Testing and Materials. ASTM STP 1443. p 131–142.

Susarla S, Bacchus ST, Harvey G, McCutcheon SC. 2000. Phytotransformations of perchlorate contaminated waters. Environ Sci Technol 21:1055–1065.

Tan K, Anderson TA, Jackson WA. 2004. Degradation kinetics of perchlorate in sediments and soils. Water Air Soil Pollut 151:245–259.

Tan K, Anderson TA, Jones MW, Smith PN, Jackson WA. 2004. Accumulation of perchlorate in aquatic and terrestrial plants at a field scale. J Environ Qual 33:1638–1646

Tan K, Jackson WA, Anderson TA, Pardue JH. 2004. Fate of perchlorate-contaminated water in upflow wetlands. Water Res 38:4173–4185.

Theodorakis CW, Rinchard J, Carr JA, Park J-W, McDaniel L, Liu F, Wages M. 2005. Thyroid endocrine disruption in stonerollers and cricket frogs from perchlorate-contaminated streams in east-central Texas. Ecotoxicology 15:31–50.

Thuett KA, Roots EH, Mitchell LP, Gentles BA, Anderson T, Kendall RJ, Smith EE. 2002a. Effects of in utero and lactational ammonium perchlorate exposure on thyroid gland histology and thyroid and sex hormones in developing deer mice (peromyscus maniculatus) through postnatal day 21. J Toxicol Environ Health A 65:2119–30.

Thuett KA, Roots EH, Mitchell LP, Gentles BA, Anderson TA, Smith EE. 2002b. In utero and lactational exposure to ammonium perchlorate in drinking water: effects on developing deer mice at postnatal day 21. J Toxicol Environ Health A 65:1061–76.

Tietge JE, Holcombe GW, Flynn KM, Kosian PA, Korte JJ, Anderson LE, Wolf DC, Degitz SJ. 2005. Metamorphic inhibition of *Xenopus laevis* by sodium perchlorate: effects on development and thyroid histology. Environ Toxicol Chem 24:926–933.

Urbansky ET. 1998. Perchlorate chemistry: Implications for analysis and remediation. Bioremed J 2:81–95.

Urbansky ET, Collette TW, Robarge WP, Hall WL, Skillen JM, Kane PF. 2001. Survey of fertilizers and related materials for perchlorate (ClO_4^-). EPA/600/R-01/049. Final Report. US Environmental Protection Agency Office of Research and Development, Cincinnati, OH, USA.

Urbansky ET, Magnuson ML, Kelty CA, Brown SK. 2000. Perchlorate uptake by salt cedar (*Tamarix ramosissima*) in the Las Vegas Wash riparian ecosystem. Sci Total Environ 256:227–232.

[USACE] United States Army Corps of Engineers. 2004. Final Report: Bosque and Leon River Watersheds Study. Available from: http://www.swf.usace.army.mil/pubdata/ppmd/Perchlorate/deliverable final.html. Accessed Jul 2004.

[USEPA] United States Environmental Protection Agency. 1997. Special report on environmental endocrine disruption: an effects assessment and analysis. February 1997. Washington (DC): USEPA Office of Research and Development. EPA/630/R-96/012. 111 p. Available from: http://www.epa.gov/ORD/WebPubs/endocrine/. Accessed 13 Jul 2006.

York RG, Barnett J Jr, Brown WR, Garman RH, Mattie DR, Dodd D. 2004. A rat neurodevelopmental evaluation of offspring, including evaluation of adult and neonatal thyroid, from mothers treated with ammonium perchlorate in drinking water. Int J Toxicol 23:191–214.

York RG, Funk KA, Girard MF, Mattie D, Strawson JE. 2003. Oral (drinking water) developmental toxicity study of ammonium perchlorate in Sprague-Dawley rats. Int J Toxicol 22:453–464.

Yu L, Canas JE, Cobb GP, Jackson WA, Anderson TA. 2004. Uptake of perchlorate in terrestrial plants. Ecotoxicol Environ Saf 58:44–49.

Abbreviations

ANOVA	analysis of variance
AP	ammonium perchlorate
ATP	adenosine triphosphate
ATR	attenuated total reflectance
BAF	bioaccumulation factor
BAT	brown adipose tissue
BD	below detection
BCF	bioconcentration factor
BMR	basal metabolic rate
CE	capillary electrophoresis
CHO	Chinese hamster ovary
CNS	central nervous system
COS	cell line derived from African green monkeys that are routinely used for transient transfections
CRF	corticotropin-releasing factor
CRLF	California red-legged frog
CSM	conceptual site model
D1	Type I 5′deiodinase
D2	Type II 5′deiodinase
D3	Type III 5 deiodinase
DoD	Department of Defense (US)
EDC	endocrine-disrupting chemical
EDSTAC	Endocrine Disruptor Screening and Testing Advisory Committee (USEPA)
ERA	ecological risk assessment
ESI	electrospray ionization
FETAX	frog embryo teratogenesis assay
FIA	flow injection analysis
FLE	forelimb emergence
FTIR	Fourier transform infrared spectroscopy
GI	gastrointestinal
GSI	gonadal somatic index

hAIT	human apical iodide transporter
HPT	hypothalamic–pituitary–thyroid
I^-	iodide
IC	ion chromatography
IDD	iodine deficiency disorder
ISE	ion-selective electrode
LC	liquid chromatography
LHAAP	Longhorn Army Ammunition Plant
LOEC	lowest-observable-effect concentration
MA	macrophage aggregate
MCL	maximum contaminant level
MCT8	monocarboxylate transporter 8
MRM	molecular reaction monitoring
MS	mass spectrometry
NAS	National Academy of Sciences (US)
NIS	sodium (dependent) iodide symporter
NOAEL	no-observable-adverse-effect level
NOEC	no-observed-efffects concentration
NRC	National Research Council (US)
NWIRP	Naval Weapons Industrial Reserve Plant
OATP	organic anion-transporting polypeptide
OECD	Organisation for Economic Co-operation and Development
OP	octylphenol
PBPK	physiologically based pharmacokinetic
PBTK	physiologically based toxicokinetic
PCBS	*p*-chlorobenzene sulfonate
PDS	Pendred syndrome gene
PND	postnatal day
PRL	prolactin
PTU	propylthiouracil
RIA	radioimmunoassay
rT_3	reverse T_3
SEM	standard error of the mean
SERDP	Strategic Environmental Research and Development Program
SNWA	Southern Nevada Water Authority
SRIF	somatotropin-release inhibiting factor
SRIH	somatotropin-release inhibiting hormone
StAR	steroidogenic acute regulatory protein

T_3	triiodiothyronine
T_4	tetraiodothyronine or thyroxine
TBG	thyroid-binding globulin
Tg	thyroglobulin
TH	thyroid hormone
TNWIRP	Texas Naval Weapons Industrial Reserve Plant
TPO	thyroid peroxidase
TR	thyroid receptor
TRH	thyrotropin-releasing hormone
TSH	thyroid-stimulating hormone
TTR	transthyretin or thyroxine-binding prealbumin
UCMR	Unregulated Contaminant Monitoring Rule (USEPA)
UCP	uncoupling protein or thermogenin
UDP-GT	uridine diphosphate glucuronosyltransferase
USACE	United States Army Corps of Engineers
USEPA	United States Environmental Protection Agency
USFWS	United States Fish and Wildlife Service
USGS	United States Geological Survey
USN	United States Navy

Index

A

B

C

D

E

F

G

H

I

L

M

N

O

P

R

S

T

U

V

SETAC

A Professional Society for Environmental Scientists and Engineers and Related Disciplines Concerned with Environmental Quality

The Society of Environmental Toxicology and Chemistry (SETAC), with offices currently in North America and Europe, is a nonprofit, professional society established to provide a forum for individuals and institutions engaged in the study of environmental problems, management and regulation of natural resources, education, research and development, and manufacturing and distribution.

Specific goals of the society are:

- Promote research, education, and training in the environmental sciences.
- Promote the systematic application of all relevant scientific disciplines to the evaluation of chemical hazards.
- Participate in the scientific interpretation of issues concerned with hazard assessment and risk analysis.
- Support the development of ecologically acceptable practices and principles.
- Provide a forum (meetings and publications) for communication among professionals in government, business, academia, and other segments of society involved in the use, protection, and management of our environment.

These goals are pursued through the conduct of numerous activities, which include:

- Hold annual meetings with study and workshop sessions, platform and poster papers, and achievement and merit awards.
- Sponsor a monthly scientific journal, a newsletter, and special technical publications.
- Provide funds for education and training through the SETAC Scholarship/Fellowship Program.
- Organize and sponsor chapters to provide a forum for the presentation of scientific data and for the interchange and study of information about local concerns.
- Provide advice and counsel to technical and nontechnical persons through a number of standing and ad hoc committees.

SETAC membership currently is composed of more than 5000 individuals from government, academia, business, and public-interest groups with technical backgrounds in chemistry, toxicology, biology, ecology, atmospheric sciences, health sciences, earth sciences, and engineering.

If you have training in these or related disciplines and are engaged in the study, use, or management of environmental resources, SETAC can fulfill your professional affiliation needs.

All members receive a newsletter highlighting environmental topics and SETAC activities, and reduced fees for the Annual Meeting and SETAC special publications.

All members except Students and Senior Active Members receive monthly issues of *Environmental Toxicology and Chemistry (ET&C)* and *Integrated Environmental Assessment and Management (IEAM),* peer-reviewed journals of the Society. Student and Senior Active Members may subscribe to the journal. Members may hold office and, with the Emeritus Members, constitute the voting membership.

If you desire further information, contact the appropriate SETAC Office.

1010 North 12th Avenue
Pensacola, Florida 32501-3367 USA
T 850 469 1500 F 850 469 9778
E setac@setac.org

Avenue de la Toison d'Or 67
B-1060 Brussels, Belgium
T 32 2 772 72 81 F 32 2 770 53 86
E setac@setaceu.org

www.setac.org

Environmental Quality Through Science®